Praise for *An Elegant Defense*

"Explains at length, and with verve, how today's fastest changing and most important branch of medical research has overturned what we think we know about ourselves and our environment. . . . Explore[s], in clear terms, the history of the research into the immune system and exactly how we understand it to work. . . . Deeply affecting. . . . A compelling modern history of—as well as an elegant defense for—the preeminent science of our time." *—Los Angeles Review of Books*

"Richtel brilliantly blurs the lines between biology primer, medical historical text and the traditional first-person patient story. . . . Richtel harnesses his reporter's eye for the human condition." *—Washington Post*

"A thorough, richly entertaining and just-wonky-enough beginner's class in immunology through the case studies of four patients. . . . These four tales help readers without prior scientific training tackle the alphabet soup of immunology. . . . Reflective. . . . Richtel marvels at the extraordinary cooperative power of the immune system. . . . The immune system remains our most 'elegant defense.' It is also our most mysterious companion." *—Wall Street Journal*

"Provocative. . . . [Tackles] one of the most complicated and vexing topics in modern medicine: the human immune system. . . . [A] warm and heart-wrenching book. . . . A remarkable tale. . . . Richtel is a gifted storyteller—he can make even dry subjects like protein signaling come alive. . . . A story about cutting-edge science, humanely told. . . . Rich and engaging. . . . By the final page, you will possess a deeper understanding of immunology and an appreciation of the ferocious battles that patients and doctors are fighting. . . . It's a thrilling payoff."
—Matt McCarthy, USA Today

"*An Elegant Defense* by Matt Richtel is one of those rare nonfiction books that transcends the genre. On one level it is a fascinating and engrossing account of the latest, and quite astonishing, discoveries involving the human immune system and how it works. But it is also a story about people facing mortality, about the passion of scientists searching for truth, and a meditation on death and how all of us struggle with the ultimate mystery. Heartfelt, moving, full of compassion, love, and human drama, this is a work of a writer of high ethical character who is grappling with big issues and deep humanistic problems. What an inspiring and wonderful read, in contrast to the sometimes glib, snarky, and sensationalistic fare offering itself as 'nonfiction.' I highly recommend this extraordinary book."

—Douglas Preston, #1 bestselling author of
The Lost City of the Monkey God: A True Story

"In this thorough investigation, Richtel details the explosion of knowledge over the past 70 years. . . . He weaves into his narrative four case studies [that] add a moving personal dimension."

—BBC, "10 Books to Read This Month"

"Matt Richtel's *An Elegant Defense* is a comprehensive and engaging primer on the body's 'ever-vigilant, omnipresent peace-keeping force.' The immune system plays an essential role in fighting infections and cancer and in regulating our normal health. Read this superb book to better understand one of the enduring mysteries of human biology."

—Sandeep Jauhar, *New York Times*
bestselling author of *Heart: A History*

"An expert examination of the immune system and recent impressive advances in treating immune diseases. . . . Richtel downplays the claims of enthusiasts who urge us to attain the strongest possible immune system. Immunity resembles less a comic-book superhero than a trigger-happy police force, equally capable of smiting villains and wreaking havoc on innocent bystanders. . . . Richtel illuminates a complex subject so well that even physicians will learn."

—*Kirkus Reviews* (starred review)

"Richtel adroitly mingles cellular biology, scientific history, medical research, and patients' experiences. . . . Ancient and intricate, highly effective and ever vigilant, your immune system is engaged in a perpetual biological balancing act. . . . Richtel approaches this essential subject with awe, his writing meticulous and empathic." —*Booklist* (starred review)

"The immune system is less war machine than peacekeeping force, seeing off viral and bacterial disruption to keep the body safe. But what if that balance shifts? Award-winning reporter Matt Richtel examines the scientific and human realities of immune anomaly through four case studies. . . . Through these harrowing accounts, Richtel interweaves the research history."
—*Nature*, "Best Science Picks"

"A deeply reported and entertainingly written exploration of the human immune system and how it works."
—*USA Today* ("5 Books Not to Miss")

"Plumbs the enormous impact of the human immune system. . . . Examines the immune system's history and future through the lens of science, records and research and the case histories of four individuals. . . . Despite the topic's staggering complexity and the extensive information in Richtel's fact-filled coverage, the heart and fine craft of a storyteller emerges most memorably throughout the book." —Mercury News

"A remarkable journey of exploration inside the human body. . . . Richly informative and engaging. . . . Eminently readable and with so many important takeaways, *An Elegant Defense* is well worth one's investment of time." —Shelf Awareness

"An engaging deep dive into our immune system."
—*Men's Health*

"Richtel's new book is so useful. . . . Give[s] lay readers a means of understanding what's known so far about the intricate biology of our immune systems." —*The Week*, "Book of the Week"

"You probably don't give your immune system enough credit. That will change after reading Matt Richtel's *An Elegant Defense*, a deeply reported account of how the immune system works. . . . The book reveals depths about both immunology and sociocultural responses to illness and disease, making the immune system's function—to maintain wellness—all the more amazing. . . . Richtel weaves dense, complex research into suspense and human drama; his book reads, at times, like a harrowing mystery novel."   —*Spirituality and Health*

"[An] entertaining survey of the science of immunology. In punchy prose, Richtel covers the history of research into the field. . . . He also provides close-ups on the case histories of four people, including two women suffering from autoimmune disorders, interwoven through discussions of myriad present-day concerns. . . . A hard-to-put-down account of the body's first line of defense."   —*Publishers Weekly*

"A sweeping overview of immunology's history. . . . The prose in *An Elegant Defense* is vibrant, conversational, direct and often funny. . . . The content is captivating and useful. . . . The writing shines."   —*Science News*

"The immense complexity and subtlety of the system makes it difficult to describe in non-technical terms to a lay audience. . . . The lives of the four patients serve to demonstrate how suppression or amplification of the immune system can be either the cause or the treatment of human disease. By weaving their histories into the book he humanizes what could have become a deeply technical piece of writing. . . . [An] entertaining and a worthwhile read."   —*The Missourian*

"Fascinating. . . . Rooted in evidence-based research. . . . Compelling."   —*AudioFile*

AN ELEGANT DEFENSE

ALSO BY MATT RICHTEL

Nonfiction

A Deadly Wandering

Fiction

Dead on Arrival
The Doomsday Equation
The Cloud
Floodgate
Devil's Plaything
Hooked

An Elegant
DEFENSE

The Extraordinary
New Science of the
Immune System

A Tale in Four Lives

MATT RICHTEL

WM

WILLIAM MORROW

An Imprint of HarperCollinsPublishers

HarperCollins books may be purchased for educational, business, or sales promotional use. For information, please email the Special Markets Department at SPsales@harpercollins.com.

A hardcover edition of this book was published in 2019 by William Morrow, an imprint of HarperCollins Publishers.

FIRST WILLIAM MORROW PAPERBACK EDITION PUBLISHED 2020.

The Library of Congress has catalogued a previous edition as follows:

Names: Richtel, Matt, author.
Title: Elegant defense : the extraordinary new science of the immune system : a tale in four lives / Matt Richtel.
Description: First edition. | New York, NY : William Morrow, [2019] | Includes index.
Identifiers: LCCN 2018038019| ISBN 9780062698537 (hardcover) | ISBN 9780062698490 (pbk.) | ISBN 9780062699107 (large print)
Subjects: LCSH: Immune system—Popular works. | Immunology—Popular works. | Immunologic diseases—Patients.
Classification: LCC QR181.7 .R53 2019 | DDC 616.07/9—dc23
LC record available at   https://lccn.loc.gov/2018038019

ISBN 978-0-06-269849-0 (pbk.)

20 21 22 23 24 LSC   10 9 8 7 6 5 4 3 2 1

For Jason and the Argonauts

CONTENTS

PART VI: HOMECOMING

AUTHOR'S NOTE

To distinguish between medical doctors and PhDs, I refer to medical doctors with the honorific Dr., whereas I refer to PhDs solely by last name. This is a painful trade-off given that PhDs not only have attained hard-earned doctorates but have made the most important discoveries in this field. I made the decision, following an informal *New York Times* style, in order to help guide the reader through a story with multiple characters, some with research expertise, tending to be PhDs, and others with clinical expertise, tending to be physicians. I beg the indulgence of the scientists, who are chief among this Odyssey's Argonauts.

Finally, I've used first names for Jason Greenstein, his family and friends, and for others whose intimacies I've shared, including Bob Hoff, Linda Segre, and Merredith Branscombe. The personal nature of their medical tales calls for more casual language.

AN ELEGANT DEFENSE

Part I

LIVES IN THE BALANCE

1

The Ties That Bind

A gray sky overhead, Jason Greenstein sat silently in the passenger seat of a Ford Windstar. It was Friday, March 13, 2015. Jason was heading to a miracle and traveling in the style to which he had become accustomed—filth.

His silver minivan, fast approaching Denver from its suburbs, looked like scrap metal on wheels. The heater coughed and spat and seemed to work only when it was hot outside. The back door didn't open. Various warning lights dotted the dashboard, alerting Jason to system failures he ignored. His maps and atlases overflowed compartments and littered the floor.

Then there was the smell. It permeated the cabin from the five-gallon gasoline jerry can he kept in back for emergencies, and from the accumulated greasy residue from endless fast-food stops. Jason could never resist 7-Eleven rotisserie hot dogs, despite referring to them as "witches' fingers" and "disgusting."

When Jason went on cross-country sales trips, which was often, he would sometimes sleep in the back of the van. He'd curl up on a stained orange Oriental rug, his head next to the gas can. Or he'd slumber on top of the boxes of shiny, bejeweled trinkets that he sold to far-flung casinos for use as promotional items.

Jason was forty-seven, with an undergraduate diploma from an elite college, graduate degrees in business and law, and no reliance on or particular reverence for those trappings. He lived from one entrepreneurial idea to another, one adventure to the next. Never was he happier than when he was driving, a dip of Skoal Fine Cut packed in his lip, rocking out to Springsteen or to a local station on the dial with some new town on the horizon. Jason was determined to discover, explore, and live his way. He was a genuine American dreamer and the van his covered wagon.

"Ma, if anything ever happens to me, I want the van taken care of. Are you listening, Ma?" he told his mother. Jason and his mother, Catherine, alternately cherished each other and fought with vicious visceral passive-aggressive dialogue that would make Arthur Miller melt.

Now Jason sat in the passenger seat, his girlfriend, Beth, driving. He was heading to pull off as unconventional a trick as even he could ever have conceived. He was determined to become a medical marvel, a poster boy, as he termed it, for a miraculous new cancer treatment. Jason was going to defy death, while standing on its precipice, one foot already over the edge.

Jason suffered from late, late-stage cancer. By any reasonable definition, it was terminal.

Fifteen pounds of Hodgkin's lymphoma was lodged in his lungs and back on the left side of his body, and it was doubling in size every few weeks. Four years of chemotherapy and radiation had ultimately failed, managing to beat back for only brief periods what is typically one of the most curable of cancers. Doctors had tried nearly everything, some drugs twice or in combinations, with brutal side effects. The malignancy always returned. Now the tumor so protruded from his back that Beth affectionately referred to Jason as Quasimodo. The mass impinged on his ulnar nerve, which left him in agonizing pain and unable to move his left hand; it was bloated and looked like a fleshy blob.

The assault on his left hand was particularly cruel. When Jason was a kid—when we were kids together—Jason was a phenomenal athlete, cunning, tenacious, a slippery-quick southpaw. He wasn't tall, but he sure could jump, an antelope with frog legs, all-state in Colorado in basketball and baseball. He had the looks to match, dark hair and dark eyes, a generous smile, his half-Italian, half-Jewish ancestry yielding an all-American mutt the girls couldn't resist. But to me, his defining characteristic was his laugh. It exploded at a high pitch bordering on soprano, often at one of his own jokes. It was sheer delight.

As Beth drove the route to Denver from Boulder, sun skirted the clouds, as if March still couldn't decide whether it was winter or spring. Jason slumped with discomfort. He wore gray sweat pants, canvas loafers, a flannel shirt—all loose-fitting because that's all he could manage to slip over the painful lumps in his body. Even his feet swelled. Jason had taken everything cancer could throw at him. His oncologist had given Jason the nickname Steel Bull because he tenaciously endured every treatment they imposed on him, often managing a joke or smile along the way.

Then, the prior Monday, at an appointment with his oncologist, Jason had received his death sentence. The doctor had examined the progression of Jason's tumor and explained tearfully that there was nothing left to do. They'd tried every treatment, all the toxic combinations. The cancer just kept roaring back. It was time to let go.

After the visit, the doctor wrote in a patient note that "the most reasonable approach, as emotionally taxing as it is, is to consider Mr. Greenstein for hospice care." He arranged a meeting with Jason's family to prepare him for palliative relief.

Further treatment, the doctor wrote, "is proving more toxic than beneficial" and would be unwarranted "unless he has a dramatic response."

Beth steered the minivan through the middle-class neighborhood around Presbyterian St. Luke's Medical Center. Jason usually loved to talk. He was a rapid-fire chatterbox. Now she could hardly get a word out of him.

After they parked, Beth held Jason by the arm as they took the elevator to the third floor. Jason had spent hours of his life in this oncology ward, sitting in a bulky tan recliner in one of the boxy rooms, enduring the noxious chemotherapy regimens. Not on this day.

Jason slowly edged himself into a chair. A nurse attached the intravenous line to the central port in his chest. First she dripped saline through to make sure the line was clean, then Benadryl to make Jason sleepy. Then the nurse swapped out those bags for another, also with clear liquid. This was something new.

Cancer is one of the world's leading killers. This is not a cancer story. Nor is it a story of heart disease or respiratory disease, accidents, stroke, Alzheimer's, diabetes, flu and pneumonia, kidney disease, stroke, HIV, or diabetes. These are the things that ail and kill us. This is not the story of any particular disease or injury. It is the story of *all* of them and the extraordinary link that binds them, the glue that defines the whole of human health and wellness. This is the story of the immune system.

It is an account of the remarkable discovery of the immune system, particularly over the last seventy years, and of the role this system plays in every facet of our health. When a scratch or cut pierces the shield of our skin, itself a first line of defense, the immune system scrambles into action. Immune cells pour in to cleanse wounds, rebuild tissue, or repair internal damage from a bump or bruise, to tend to burns and bites. The complex defense network of cells attacks each cold virus—two to three a year—surveys the countless malignancies that threaten to become cancer, holds in check viruses like herpes that colonize huge swaths of the population, and confronts hundreds of millions of cases each year of food poisoning. Only recently have we begun to understand the pervasive role of our immune system in the brain, where damaged or outdated synapses get pruned by the organ's own immune cells, allowing ongoing neurological health.

This vigilance is constant and largely invisible to us, with the immune system a literal bodyguard that defines health in the broadest possible terms. For instance, the very mechanisms that defend our individual health appear to play a role in such essential functions as how we pick mates—helping us to avoid incestuous pairings that might damage our collective security and survival.

The immune system is often described with the language of war, one that pits our internal forces against evil disease by using powerful cells capable of surveillance and spying, surgical strikes and nuclear attacks. To expand on the war metaphor, our defense network relies too on covert agents equipped with suicide pills, and it is connected by one of the world's most complex and instantaneous telecommunications networks. This defense apparatus also enjoys a status virtually unrivaled by any other facet of human biology. It roams the body freely, moving through and across organ systems. Like police in a time of martial law, the immune system seeks out threats and keeps them from doing mortal harm, ably discerning up to a billion different alien hazards, even ones not yet discovered by science.

This is an extraordinarily complex calling, given that life is a raucous festival, your body like a sprawling party, a chaotic and exuberant affair populated with a variety of cells. There are billions of them, tissue cells and blood cells, proteins and molecules and invading microbes.

The immune system's policing job gets complicated by the porous nature of our bodies' borders. Just about every organism that wants to get inside us can do so. Our body is a take-all-comers bash, a festival with open seating, coursing with every life-form that happens by—petty thieves and gangs; terrorists armed with nuclear suitcases; dumb drunk cousins and relatives; enemy agents cloaked as friends; and foes so unpredictable and alien that they seem as if beamed from another universe.

And yet, for all these threats, the war metaphor is misleading, incomplete—even arguably dead wrong. Your immune system isn't a war machine. It's a peacekeeping force that more than anything else seeks to create harmony. The job of the immune system is to circulate through this wild party, keeping an eye out for troublemakers and then—this is key—tossing out bad guys while doing as little damage to other cells as possible. This is not just because we don't want to hurt our own tissue. It is also because we need many of the alien organisms that live on and in us, including the billions of bacteria that live in our guts. A convincing argument is now being made that some of these microbes, far from threatening us, are welcome as essential allies. Our health depends on our harmonious interaction with a multitude of bacteria. In fact, when we use antibiotics or antibacterial soaps or encounter toxins that harm our gut flora, we risk impairing bacteria that contribute to the effectiveness of our immune system's function.

And when our immune system gets overheated, look out.

Like an out-of-control police state, an unchecked immune system can grow so zealous that it turns as dangerous as any foreign disease. This is called autoimmunity. It is on the rise. Fully 20 percent of the American population, or 50 million Americans, develops an autoimmune disorder. By some estimates, 75 percent are women, with conditions like rheumatoid arthritis, lupus, Crohn's disease, and irritable bowel syndrome (IBS)—each terrible, frustrating, debilitating, hard to diagnose. Together, autoimmunity is the third most common disease category in the United States (after cardiovascular disorders and cancer). Diabetes, the leading killer in the country, is caused by the immune system's going to war against the pancreas.

The last few decades of immunology, the science of the immune system, have taught us about another core facet of the immune system: It can be duped. Sometimes a disease takes root and starts to grow and spread and then tricks the immune system into thinking it isn't so bad after all. It deceives the entire defense system into helping it grow. This is what happened to Jason.

Cancer played a nasty trick on his elegant defense. It overtook the immune system's communication channels and instructed his body's soldiers to stand down. Then it used his immune system to protect the cancer as if it were a precious, healthy new tissue, and it would have sent him spiraling to his grave.

The clear liquid that dripped into Jason's chest that auspicious Friday the thirteenth was aimed at reversing cancer's trick. It was instructing his immune system to fight. Jason, one of the first fifty patients to try one of the greatest developments in the history of medicine, was now as much a hyperkinetic frontiersman as he ever dared dream. He stood at the very edge of human achievement as modern science challenged one of the most enduring and effective killing techniques in the pantheon of disease.

When it became clear that Jason just might exemplify a remarkable shift in medicine, I picked up my pen.

As a *New York Times* journalist but also as Jason's friend, I began a journey to understand the immune system, how we'd gotten

The author with Jason Greenstein. "I'm back," said Greenie. (Nick Cote/New York Times)

to this place where we might tinker with it, and what that means. What I found was a story of scientific discovery and heroics, a global detective yarn that spins through Europe, Russia, Japan, and the United States, with researchers stacking one hard-earned revelation on the next. The sum of my learning is a series of pivotal vignettes and lessons, personal histories and scientific aha moments, making this book less textbook than tale. It's the story of the immune system's mechanics and its reach into the practical side of health—into sleep, fitness, mood, nutrition, aging, and dementia.

It's also the story of Jason and three other medical wonders you'll meet: Bob Hoff, a man with one of the most unusual immune systems in the world, and Linda Segre and Merredith Brans-combe, two stalwart souls fighting an invisible killer—their own overactive immune systems.

Like Jason, they are part of a monumental scientific inflection point, an explosion of knowledge, with experts placing our new understanding of the immune system on par with the greatest human achievements.

These new discoveries are "as significant as the discovery of antibiotics," said Dr. John Timmerman at UCLA, who has done pioneering research into the immune system. In terms of fighting a host of diseases impacting both quality of life and longevity, "we're like Apollo 11 right now. We have touched down. The *Eagle* has landed."

At St. Luke's that Friday the thirteenth, the drug dripped into Jason's system for an hour, after which Beth drove them forty-five minutes back to Boulder, where he planned to watch his nephew Jack's high school basketball game at the Coors Events Center on the campus of the University of Colorado. When they arrived at the game, Jason lacked the strength to climb the arena's stairs, so a family member talked the event staff into letting him in through a special entrance and directly onto the court.

This was how Jason used to go into games in his heyday, right onto the court, the center of it all. In fact, at this very arena, decades earlier, I'd sat in these stands and watched Jason drill one of the most improbable, thrilling shots I'm ever likely to witness again in my life. His winning shot from the top of the key came at the buzzer of a double-overtime game against a rival, allowing his team to advance in the state play-offs.

Many years later, Jason sat in the stands as friends wandered over and looked at a dwindled shell who they felt certain was attending his last game.

"He looked so bad," observed one of Jason's old friends and teammates, a skinny sharpshooter named Danny Gallagher. "I wondered if he'd make it through the night."

2

Jason

The immune system story is one of life and death, of course, a story of survival under the deadliest conditions. As much as that, it is about the struggle for peace and harmony, for successful integration, immigration of organisms across body and border, manifest destiny, and evolution. It is a story of friendship.

My earliest memories of Jason were in the infield and dugout. We were ten. Our Little League team was sponsored by McDonald's. White uniforms with yellow trim. Jason had a puff of big hair and a smile nearly as wide. In team pictures, he stood in the back row. I knelt in the front, happy in many ways and confident in school, but hiding the growing insecurity of a short kid craving attention.

Jason seemed to me to embody the ideal, an all-American boy, not just a great jock, but blessed as well with a natural curiosity, kindness, and mountains of charisma. In seventh grade, he was voted most outstanding student. When he was present, others yielded to him. He was nicknamed Golden. He was more appealing a person because he was the opposite of a bully. "Go get it, Rick!" he'd shout when I'd be at bat, likely to strike out, lucky to walk. "Next time," he'd tell me when I got back to the bench.

We had some things in common, notably fathers we looked up to, who loomed large in our lives and in the community. My dad

Jason, second from left on top; author, bottom right, below Jason's dad. (Courtesy of the author)

was a judge in our relatively small town. Jason's dad, Joel Green-stein, was a beloved divorce attorney, our Little League coach, *the* Little League coach in town, our own version of Walter Matthau, without the cursing or drinking. He chomped a stogie, had a wry smile and dry wit, was visible a ball field away in his navy blue Yankees windbreaker. He'd stand in the dugout with his knee up on a step, fist smacking into his cracked-leather catcher's mitt.

Joel doted on and gently but strategically steered Jason, like a judicious trainer who had lucked into a Thoroughbred.

"Jason adored my father," Jason's sister Yvette told me. "He was so close to my dad, and my dad just adored him. My dad was a more reserved kind of person, and here was Jason, who always had it all out there, no filter, really, emotionally; he gave you whatever he had at the time."

Guy, Jason's older brother, said of Jason: "My dad was his guru."

From a health perspective, there was a powerful material difference between our fathers—Murray (my dad) and Joel. Murray discovered jogging at the inception of the 1970s running craze and became as fanatical as anyone, eventually completing thirteen marathons. Joel was fit too, but he smoked cigars. Jason's mother, Cathy, smoked a pack of cigarettes a day. Tobacco smells permeated the Greenstein house. Smoking tests the immune system like few human habits; the tiny nicks and cuts to the soft lung tissue don't just create persistent injury but force cells to divide to replace the hurt tissue. Cell division heightens the possibility for malignancy, cancer. This is just simple math, and it can be deadly.

In eighth grade, Tom Meier, one of Jason's closest friends, was standing in the school gym. The door flung open and Golden rushed in. "He was sobbing," Tom recalled.

Before Tom could fully get Jason's attention, Jason made his way to the locker room, and Tom followed. Jason sat on the locker room bench.

"What's up, J?"

"My father is dying."

Jason had learned his father had colon cancer.

Forty years later, Tom tears up as he tells the story. "Here was the strongest person I knew in the whole world," Tom told me, "and he was absolutely shattered."

Jason seemed outwardly impervious to the malignancy eating his father alive, the victim of his own growing emotional disconnect. In ninth grade, Jason ran for president of the student council. His speech exuded confidence and grace. He told the school he would never give up.

"If elected, I will try to be consistently committed and not lose my spice or vigor." He had but one promise. "I will do my best and try my very hardest for you if elected president."

If. Of course he won.

Then while we were in tenth grade at Boulder High School, Jason came up with the life philosophy that would define us all for a few naive and wonderful years. He gave a name to a group of friends: the Concerned Fellows League. The CFL.

It was a world view adopted as an organizing principle by Jason and six others of us who were a tight-knit group in high school— Josh, Noel, Tom, Adam, Bob, Jason, and me. The CFL philosophy essentially was the opposite of what it seemed to express. Jason's point was that we were not particularly concerned. Worrying was for people who had lost perspective.

Like all of life's enduring philosophies and religions, this macho idea folds in on itself and becomes a complete contradiction. Don't look too closely. We were, to a person, worried about all kinds of things, scared as hell and insecure, despite our privilege. This kind of disconnect, as you'll see, can lead to anxiety, and illness, all of it connected to the way the immune system deals with stress. But outwardly, at the time, we were a combination of the fortunate good students and athletes, the "cool crowd." Jason carried the torch. In eleventh grade, he did something remarkable.

As an undersized underclassman, he helped the Boulder High School Panthers on a magical run through the 1984 basketball state play-offs. He was five-nine in high-tops, not the lone star on the team—there were several top-notch seniors—but Jason arguably was the glue, point guard, and mascot, unmatched in intensity.

His high school coach of that championship-game team, a Bobby Knight–like figure named John Raynor, thought of Jason as the kid who couldn't be broken. "He played at times with reckless abandon," Coach Raynor recollected. He'd dive to the floor "and come up lame, and I'd think, Gosh, is this guy going to survive?"

In the stands at the championship game—for bragging rights of the whole darn state—the members of the CFL sat and cheered, our faces painted with the little purple paws of our Boulder Panthers.

Sitting not far away from us, a shrunken shadow holding on for dear life, Joel watched his beloved son.

The game went badly from the start.

Jason, already outmatched in size and strength, labored with a wobbly ankle he'd hurt in the previous game. He scored only four points. The Panthers' two great shooters were shaky. Final score: 52–42.

Just a few months later, on July 13, 1984, Joel Greenstein died. He was fifty.

Jason got the news and came home from work to find his father laid out on a stretcher in the living room, being attended by hospice. He sobbed. Jason somehow didn't believe this could happen.

Jason would later tell me, "There are two things I hate in this world, hospitals and cancer."

Some in his family wondered whether the death of his father, his ballast, left Jason so unmoored that he went on the run—physically, spiritually, emotionally. After Joel's death, Jason ran harder and faster, a racehorse without his trainer. It was a full-tilt lifestyle, world travel—teaching in Japan, traipsing through

Latin America—and multiple graduate degrees. Sort of. His fees unpaid, he could never get the law school diploma he'd earned. He became a serial entrepreneur and one-man sales band, selling mobile phone service, Crocs at the mall, juicers to restaurants. He built and ran a ski van company. Each of his ideas was devised with the enthusiasm of a guy drawing up the winning shot.

It would seem, looking back, that he was putting his health at risk, but it was I who had the first brush with illness. After college, I broke down, succumbing to the pressure of inflated, misguided ambitions without any real clue as to my true passions. Insomnia and anxiety came too. I had to find myself to survive. Out of the process I emerged as a person largely content in my own skin and suddenly able to follow my own muse without fear.

By the late nineties, I, healed and happy, and Jason, adventurous and coming up with one business idea crazier than the next, formed a deep, authentic friendship. Enthusiasm bonded us, and old times, along with an ability to simultaneously not take ourselves too seriously while becoming consumed by the righteousness of our respective muses. Then fate arrived for Jason.

He landed into a Phoenix airport under a gorgeous evening sky on May 9, 2010. It was a Sunday night and Jason had spent the weekend at a gambling industry trade show in Biloxi, Mississippi. His latest business entailed selling Chinese-made trinkets—small decorative enamel boxes—to casinos to give as prizes to loyal customers or to instant winners. The company name was Green Man Group.

It was as Jason as Jason could get. He lived in Vegas, the frontier of the gambling man, selling shiny things to fellow dreamers, and traveling the country to visit the growing numbers of casinos to ingratiate himself and explain why his trinkets would send their customer loyalty soaring. He drove a 1982 Chrysler Concorde, which he described to me as "98 percent of Jews' last cars. Every one of those Jews died or couldn't drive or sold the car to a Mexican family. And every one of those was owned by a Mexican family except the one that I'm driving."

Then he squealed his high-pitched laugh, either with self-awareness at the slightly off-color comment or maybe the opposite—he just thought he was funny. And it was nearly impossible not to laugh along with him. This was Jason in his element, the windows down, the air warm, an adventure ahead. "I loved driving in the desert and being in the open road."

He'd stopped in Phoenix on his way home to Vegas because he had some business in Arizona. When he landed, late on the ninth, the airline had misplaced his luggage, which included his sample kits of trinkets. He had to stand there, waiting. He felt a tickle in his throat. He thought, *Sometimes I get allergies in the desert, or I've got strep throat or a virus.*

He stayed half an hour from the airport at a Days Inn and felt crummy in the morning. It bummed him out. "It was a beautiful day in May, and I felt really yucky, had a headache." To pep himself up, he did what he always did when he drove, popped into his lip a plug of Skoal Fine Cut chewing tobacco—"I chewed like a madman"—and then, when he still felt cruddy, stopped for a snack at a gas station.

There he was, feeling like shit, on the open road, which was his place, his happy place.

"Jason was one of those people who would've settled the west," his sister Natalie described him. "He'd have left the city and risked the Indians or whatever." She wasn't sure if that was just his makeup or if his constitution was amplified by the death of his father and "when our dad died, something in him broke or switched." Settling down, slowing down—that just wasn't how Jason rolled. He had his own ideas and pursued them when others would think them wide-eyed in the extreme, just like the home-grown treatment he came up with a few weeks later to cure his sore throat.

Jason lived in Vegas with—who else?—a stripper. She rented a room from him in the house that his mother bought for him, as an investment, for $175,000. It was a ranch-style three-bedroom with a pool in back, built in 1947, and it was sometime after that—but long before the Greensteins bought it—that this neighborhood had its heyday. At one point, a casino magnate had lived across the street, and Jason planned on renovating and flipping the house. That's what he said.

The relationship with the stripper was strictly platonic. That was mostly fine with Jason. Also, Jason had Beth.

The Friday after he'd first gotten sick, he still couldn't kick the symptoms. "I did what most people would do," he said, laughing. "I went out late Friday and bought a case of beer and got hammered to try to get rid of the cold."

Jason woke up the next morning feeling worse. "I tried drinking it out of me, and that didn't work very well."

He called Beth, and she told him, "You've got to go to the doc-tor.'" He went, and they did a blood test and noticed a large raised lymph node on his neck. The doctor thought he had mono and gave him antibiotics. The drugs didn't work.

"I felt no difference whatsoever."

Every summer, Jason drove east with his mother back to New York to see her family. She hated to fly. And she and Jason had a kind of codependence and mutual affection that was easy to mistake for professional wrestling, at least of the verbal kind. They would bicker, voices rising histrionically.

Ma, you're not listening! I don't feel well.

Jason, if you're not feeling well, go to bed!

I'm fine, Ma. I'm driving you to New York.

That's nice, Jason. That's sweet of you.

He drove to Colorado, picked her up, and they went east. *I'm really weak,* he thought. It was mid-June when they arrived in Bayside, Queens, their annual manifest destiny trek in reverse that brought Jason back to the family's American point of origin. There, at his aunt Rose's house, he could not get off the couch.

"It reminded me of my dad when he got sick. He had never done that before," Jason recalled.

Jason didn't have a regular doctor. In fact, he didn't have proper health insurance.

"I'd recently purchased a bogus health care insurance policy online. It said it was an emergency policy. Cancer didn't count. It only paid up to a thousand dollars. That was my lifestyle—like betting your tenant a bottle of Captain Morgan that her tits were real."

Back in Colorado, he finally got a blood work-up. One of the tests measured inflammation through a nonspecific examination of his erythrocyte sedimentation rate. Jason's number was off the charts.

The doctor called Jason. "We've got a real problem here." He recounted Jason's test results. "I've never seen anything like this in my thirty years. Something is very wrong."

Jason was diagnosed with Hodgkin's lymphoma. His immune system was being swamped by malignant forces. On the bright side, Hodgkin's was among the most curable of cancers—for most people.

3

Bob

Robert T. Hoff became an immune system marvel on Halloween night of 1977. He was dressed as a mummy.

Born in 1948, raised in Iowa, son of an insurance man and a substitute teacher, he'd been in hiding since he was four. That was the first time he could remember that he and a boy next door fondled each other. He loved it, craved physical affection from other boys and eventually from men. He learned to hide the reality that for a few early years he dressed in his mom's dresses and scarves. He overachieved in school. He didn't tell anybody about his passions after he made the mistake of doing so once, in seventh grade, and the kid he told, Steve Lyons, blabbed it.

"I was referred to as a flaming faggot."

Bob needed a new strategy. He found it in imitation. There was a kid named Art who was the most popular guy in school. Bob learned to emulate Art.

"I picked up on everything he did. I picked out his extracurricular activities. I swam at the YMCA. I learned how to speak differently. There's a gay accent, and I learned how to think in advance to not use a word that would contain a lisp.

"Then I started to become popular, the star of the school play, elected president of the student government, the most popular kid in class."

He dated girls and stopped having sex with men until college, fearing he'd be ostracized.

He went to law school and married a woman. He went on active duty in the Air Force. He and his wife tried to make it work. She didn't want to be married to a homosexual. They divorced. He married again. At some point, Bob's mother found out about his true proclivities. They didn't speak for twenty years because she thought him sinful.

By 1977, Bob lived in Washington, D.C., now an accomplished lawyer—assistant general counsel for a major federal office, the General Services Administration. On October 31, Bob went alone to a party; his wife at the time, a flight attendant, part of the cover of his life, was out of town.

Bob wrapped himself in gauze he'd bought in a thirty-foot roll at Joann Fabrics and was hanging out at the party when he met John. John was a very fit redhead. Bob and John went upstairs and had unprotected sex.

Two weeks later, Bob felt dizzy, lethargic, and tired, with achy flulike symptoms—not enough to keep him from work. The discomfort lasted ten days. "I chalked it up to the flu," Bob recalled.

Around Thanksgiving, Bob went to his cousin's wedding in Cedar Falls. On the drive back, he felt really sick. He threw up and had diarrhea. He assumed he'd eaten bad shrimp. Bob, an overachiever all his life, went to see the doctor who had given him a physical when he applied for his private commercial flying license.

Bob had hepatitis. The version he had was hepatitis A, a strain that had been identified only a few years earlier, in 1973. It is an infection of the liver that takes some time to manifest. When it does, the symptoms that a person—in this case, Bob—experiences typically are those that come from the work that the immune system is doing when it fights back, inflammation.

For Bob, this diagnosis wasn't such bad news, all things considered. If the immune system does its job properly, hepatitis A is a strain that can be overcome.

But this wasn't all Bob had. Bob had contracted the human immunodeficiency virus, HIV, arguably the most serious direct threat to ever confront our immune system. It would take a few years before Bob would discover the truth. Then he would become a source of powerful inspiration and wisdom at the highest reaches of the scientific world. In the medical realm, Robert Hoff is a veritable state treasure. His body parried HIV, and death, as perhaps no one had done before him, so his precious immune system offered insights and promise for the rest of us too.

4

Linda and Merredith

There was little to suggest that Linda Bowman harbored inside her an invisible suicidal assassin as she stood on first tee of a wind- and rain-plagued golf course in Ulster, Ireland. It was May of 1982, during the final round of the Smirnoff Ulster Open, a precursor to the Ladies Irish Open. Linda was tied for the lead.

Just prior to her two P.M. tee time, her gruff, standoffish local caddy, Victor McCauley, had stunned her by taking her to the parking lot. "I've got something to show you," he said. He opened the trunk to reveal a dozen beautiful red roses. "Linda," he said, "let's go win this thing."

It wasn't going to be easy. Linda, twenty-two, had never won a professional golf tournament, and she was up against a woman who was the leading money winner for two years running on the European tour. She'd hardly slept the night before the final round.

On the other hand, much about Linda's life had been storybook—*storybook,* not fairy tale. She hadn't been handed everything, wasn't a princess. She worked her tail off, and she liked to work. As a girl, starting at age seven, she'd poured herself into horseback riding, reaching competitive heights. She pushed herself, looked for her edges, even to the point of occasionally putting herself while in her early teens on protein-only diets— meat and eggs and no fruits or vegetables—to stay lean and grace- ful on the mare.

She became the best competitor in the barn. "You could give me a terrible horse, and I could make it perform."

And brains? Linda, excelling in particular in math, emulating her older sister, had skipped third grade.

She was well liked, maybe not the most popular, a touch nerdy, but content and self-motivated. Her mom had been a professional golfer, her dad a scratch golfer too, and eventually Linda moved from horseback riding to the family business. From almost a standing start at age fifteen, she worked relentlessly at golf and elbow-greased herself a Stanford University golf scholarship. Drives flew 230 yards, a feat in those days.

In the final round of the Ulster Open that May day in 1982, Linda kept even with the tour leader, Jenny Lee Smith. On 18, the last hole, Linda scrambled after her second shot, which trickled over the green into a bunker. Her sand blast landed six inches from the cup, and her par sent the match into a sudden death play-off.

The two players went head-to-head, one hole at a time. If one of them won a hole, a single hole, she would take all the marbles. Linda and Jenny played even for four holes, and then, on the fifth, a 500-yard par five, Linda cracked. After the two ladies hit nearly identical solid drives, Linda pulled out her three-wood, swung, and. . . . topped the ball. It lazily spun on the ground 90 yards, well less than half the distance she'd hoped for. Jenny stood up next and crushed it. Jenny had only to hit a decent wedge shot and the tournament would be hers.

Victor, Linda's rose-bearing caddy, calmly handed Linda a five-iron and told her she knew what to do. Swing gracefully, with power and confidence, and put the ball near the pin, stay in the match, put pressure on her opponent.

Linda struck her five-iron "right on the stick," and the ball landed three feet from the hole. Jenny hit her wedge shot over the green. When Linda made her birdie putt and won, her American teammates put her on their shoulders. Later, in after-party celebration, she danced with that old caddy to "Forty Shades of Green."

Linda Bowman had many gifts, chief among them work ethic, grace under pressure.

Until her body turned on her.

Fourteen years later, in 1996, it seemed from the outside that much of Linda's life had continued in storybook fashion. She had a Stanford MBA to go with everything else, two kids, including a newborn, a husband employed at one of Silicon Valley's most elite law firms. She was on the cusp of becoming only the sixth female partner at the Boston Consulting Group.

She lived in a nice home in San Mateo, a suburb of San Francisco. One night in September of that year, she was cooking dinner for a group of colleagues when she felt pain in her left big toe. Not just a twinge. PAIN! She looked and saw that the toe had swollen to the size of a golf ball. In agony, she powered through dinner and then took a rare step for such an overachiever and politely asked her dinner guests to leave early.

Then, even more uncharacteristically, she cancelled the meeting she had scheduled for the next day. She had been supposed to fly to Los Angeles to meet with a major client, one of the world's largest banks. But she couldn't imagine herself getting to and through the airport.

To go to sleep, she took a Vicodin she had left over from the birth of her son. The pill didn't work. She took a second. No relief. She swallowed a third.

The next day or so, she went to the doctor and unveiled a toe that was the size of the Titleist balls she once crushed. It was red and puffy, a balloon of excruciating pain.

Linda's doctor examined her. "I'm not sure what this is," the doctor said.

Linda was under attack by her own body. She suffers from rheumatoid arthritis. Her story will sound familiar to legions dealing with autoimmunity. She has dealt with terrible pain and swelling—swelling of one thing or another, bowels, organs, and joints in particular.

Broadly, it is hard to overstate the toll taken by autoimmunity. Of the five top-selling drugs on the market, three treat autoimmunity, including the world's bestselling medication, Humira, used to suppress the immune system for treatment in a variety of illnesses. It boasts nearly $20 billion in annual sales.

For all the sufferers of autoimmune disorders, the medicines show how far science has come in treating these conditions and understanding them. What we can see clearly now is that arthritis sufferers, people with celiac disease or lupus, even people who suffer seemingly mysterious bouts of fatigue, fever, and pain, all share an often-invisible threat: an elegant defense that is out of balance, an immune system that has overcompensated, been triggered to act without proper constraint. These conditions impact millions—well more than get diagnosed—whose own defenders attack or reject themselves, and sometimes their food or environment, as if it were hostile.

Linda's story provides one intimate window into the way autoimmunity plays out, not just the physical agony but the often unending frustration involved in trying to diagnose these complex medical conditions.

That frustration is underscored with the story of a second autoimmune sufferer, Merredith Branscombe. Her condition made her feel invisible in the sense that there is no foreign agent inside to identify, just Merredith. For decades, sufferers like Linda and Merredith have been overlooked, even dismissed by many friends, family, even medicine.

In Linda's case, the clues to and the catalyst of her illness were actually there all along to be discovered if given proper scrutiny. In addition to her family history, she suffered extreme stress, sleeplessness, and a case of strep throat that might have kicked her immune system into overdrive. Merredith's case proved more vexing.

Merredith was born in Denver, just two years after Linda, to a world of autoimmune minefields. Her family held a great secret she didn't learn for years. Her grandparents and mother had escaped in harrowing fashion from the Nazis. It added trauma to a stark family history of odd symptoms, such as the fatigue and gastrointestinal struggles of her mother, and her grandfather's own rare autoimmune disorder that attacked his nervous system.

Merredith was a good student with politically active parents, her father a newspaperman, and she grew up a gifted writer. But she was plagued occasionally with odd physical symptoms—rashes, stomach problems, joint pain—that came and went. Life seemed good when she got into Northwestern University. Then, during her junior year she was sexually assaulted and returned home from college. She was an immune system tinderbox.

When her condition exploded, it was truly something to witness.

One day in September of 2017, I met with Merredith in Colorado. It was just after five in the afternoon, and Merredith stepped out of her beige Toyota, looking very much out of place. The temperature had just dipped below 80 and the sun was still piercing even at this hour, particularly at a mile above sea level. But Merredith, fifty-three, wore jeans, a long-sleeved black shirt, and a black baseball cap, her full blond hair spilling out onto her shoulders.

She opened the back door of her aging Camry and out jumped Bam-Bam and Ringo, two mutts with hounds' blood in their genes.

We were in Boulder, my hometown, and largely coincidentally, the place where Jason and I had grown up. As Merredith leashed the eager dogs, I began to assimilate her seemingly odd choice of clothing. Of course, I thought, it has to do with her ailment. Rather, ailments. Plural.

Merredith had been diagnosed with at least three autoimmune disorders, including lupus and rheumatoid arthritis. Merredith's immune system had turned on her own body as if it were itself an alien threat. She was almost never without a challenge, often running a low-grade fever twenty days or more every month, a few days of that as high as 100. It was just enough to create regular fatigue, not enough to knock her out completely. When the symptoms hit hard, "Ouf," she said. There were middle-of-the-night emergency room visits with inflammation around her heart, blood in her stool, and pain "like someone had plunged knives into both sides of my body and were just . . . turning and driving those knives deeper and deeper into my muscles."

She shut the Toyota's back door. "Want to see something cool?" she asked.

"Sure."

"I'm going to show you what happens when I get into the sun."

I was pretty sure the thing she was going to show me was not cool. It was going to be riveting, maybe, or instructive about the power of the immune system. Not cool, not if you're Merredith.

"It's a little heartbreaking, because in general I have a fair amount invested in an image of myself as not stoic exactly, but I don't want to be the person for whom being sick is the biggest thing that ever happened to her," she told me.

Dogs leading the way, we walked up a street called Linden Avenue toward the foothills. As we moved past the tree cover, we reached a dirt walking path, the mountains and yellow-orange sun to our left, and to our right, the tree cover of an affluent neighborhood. For the moment, we were exposed.

"Check this out," Merredith said. She pulled her black shirt over her left hand, protecting it from the sun. She held her right hand out in front of me, palm down. "It's going to happen fast."

"What is?"

"Just watch."

The uncovered hand began to swell. It turned red.

"Are you okay?"

"Eh." It seemed this was par for the course.

"Let's get out of the sun," I said.

We walked another ten yards.

"There it is," she said. She withdrew her left hand from her shirt, and put the pair of them side by side. Now it was more glaring, her left hand white and a touch puffy, which reflected the regular inflammation, her right hand red and visibly swollen.

"My immune system," she said, "is always attacking me."

Merredith's immune system is far out of whack, unrestrained, a killer inside her. So is Linda's. With Jason, his immune system didn't do enough, not on its own. In the case of Bob Hoff, his immune system accomplished the rarest of feats. It was a marvel. So why was he so shunned by society?

In combination, theirs comprise a kind of immunological Goldilocks story: Two people had too powerful an immune system, one had too weak a system, and one's system was just right.

In these pages are their stories and other intimate medical tales, including some from leading scientists and, in a spot, my own health struggles. The personal accounts bring to life the powerful and complex immune system science.

What is happening inside our bodies will make more sense if I start at the beginning, with how scientists came to understand the true meaning of the immune system, and return in detail to the stories of Jason, Bob, Linda, and Merredith.

It's a tale that begins with a bird, a dog, and a starfish.

Part II

THE IMMUNE SYSTEM AND THE FESTIVAL OF LIFE

5

The Bird, Dog, Starfish, and Magic Bullet

A case can be made that the field of immunology originated with a chicken.

The setting was the University of Padova in northern Italy, at the end of the sixteenth century. There was at that time a young researcher named Fabricius ab Aquapendente who liked to cut things up. He dissected eyes, ears, animal fetuses, and occasionally humans. But history remembers him for a chicken.

Dissecting a fowl one day, Fabricius noticed an odd region beneath the chicken's tail. He found a saclike organ, which he called the bursa, a word that shares its derivation with the modern word *purse*. Henceforth, the bursa of Fabricius.

This thing appeared to have no purpose. What the heck was it? Why would God (this was the sixteenth century) leave on a bird a saclike purse that didn't seem to do anything?

Would Fabricius have believed it held the key to understanding our survival? Could he have known this simple observation would someday save the lives of millions of people, including Jason's?

So too it would be for a handful of other seemingly unconnected discoveries that would build the foundation of our understanding of our immune system.

On July 23, 1622, an Italian scientist named Gaspare Aselli dissected a "living well-fed dog," recounts one history of this seminal surgery. In its stomach, he observed "milky veins." This observation wasn't consistent with an understanding of a circulatory system carrying red blood. Instead, these milky veins looked like they contained white blood. Aselli's dissection set off a period of exploration that the history calls lymphomania, a fascination with a little understood bodily fluid called lymph, along with the dissection and vivisection of hundreds of animals.

The role of the milky veins wasn't clear for many years. As *Nature* magazine put it centuries later, Aselli's observation "languished in relative obscurity for decades."

What was this alternate circulatory system?

In northeast Sicily in the summer of 1882, Élie Metchnikoff peered through a microscope. Metchnikoff was a zoologist from Odessa who had gone to Italy to visit with his sister and her family during a period when trouble was brewing in Russia. Jewish farmers faced intensifying persecution from the government and peasants. At one point, the peasants murdered a farmer. Metchnikoff took his microscope to Sicily, where lightning struck: "The great event of my scientific life took place."

The name Fabricius will be forever linked to the chicken bursa. For Metchnikoff, the association is with a starfish larva. This was the medium of his great observation.

One day while his family was at the circus—"to see some extraordinary performing apes"—Metchnikoff turned his microscope on embryonic starfish, which are transparent. He noticed cells moving throughout the tiny organisms. He described them

Élie Metchnikoff was years ahead of his time in observing immune cells. (Wellcome Collection)

as "wandering cells," and he was struck by an instant of revelation.

"A new thought suddenly flashed across my brain. It struck me that similar cells might serve in the defence of the organism against an intruder," he wrote.

He had an idea how to figure it out. What if, he wondered, he put a splinter into a starfish? Would cells like these somehow swarm, as if coming to the rescue?

> There was a small garden to our dwelling, in which we had a few days previously organised a "Christmas tree" for the children on a little tangerine tree; I fetched from it a few rose thorns and introduced them at once under the skin of some beautiful star-fish larvæ as transparent as water.

> I was too excited to sleep that night in the expectation of the result of my experiment, and very early the next morning I ascertained that it had fully succeeded.

Indeed, a bunch of these wandering cells swarmed around the splinter. They appeared to eat away at the offending or troubled tissue.

> That experiment formed the basis of the phagocyte theory, to the development of which I devoted the next twenty-five years of my life.

The word *phagocyte* is taken from the Greek and can be roughly translated as "devourer of cells."

Phagocytosis is the process by which the devouring happens. (And congratulations, reader! You've been introduced to the language of immunology, at times one of the most maddening and even counterintuitive lexicons ever contrived.)

Metchnikoff's sister wrote his biography and eloquently captured her brother's theory, one that took years for scientists to fully embrace. "This very simple experiment struck Metchnikoff with its intimate similarity to the phenomenon which takes place in the formation of pus," she wrote, as passing of cells leads to "inflammation in man and higher animals." In the biography, she defined inflammation as "a *curative reaction* of the organism, and morbid

Paul Ehrlich, immunology godfather, in his lab. (Wellcome Collection)

symptoms are no other than the signs of the struggle between the mesodermic cells and the microbes."

In other words: At the moment of invasion, the body has an initial reaction that involves the swarming of eater cells, and the experience is not always pleasant. This is what we call inflammation.

Know this about Metchnikoff: The man was *way* ahead of his time.

Nine years later, in 1891, a contemporary of Metchnikoff's named Paul Ehrlich—a godfather of immunology who was based in Berlin—began a search for a "magic bullet." Dr. Ehrlich aimed to explain one of the most elusive questions in all of immunology: How was it that our defense system could recognize and attack dangerous foreign agents, called pathogens, organisms like viruses, bacteria, and parasites? How did the cells in the body of the starfish, for example, know to show up and start eating?

He was building on a personal obsession with a scientific technique that allowed tissue to be stained. In this way, he could see that some chemicals have "a marked affinity" for certain parts of the body, explains a history in the journal *Pharmacology*. For instance, the history notes, the chemical methylene blue seemed to travel to the nervous system. Or was the nervous system attracting the chemical?

Was there a magic bullet or some other substance or process that would allow a defense cell to attack a bad guy?

The breadth of the answer would elude scientists for years. The question was the right one.

Dr. Ehrlich had a theory. It was both brilliant and wrong. He thought that maybe the human defense system was built around a lock-and-key mechanism. When a disease came along, special cells of the body would come into contact with and attach to the virus or bacteria. Dr. Ehrlich gave a name to the attachment. He called it *Antikörper*. In English: antibody.

The idea was that antibodies would attach to parts of the disease called antigens. The antibody was the key and the antigen was the lock. Then the antibodies would help destroy the cell. There were a few problems with Dr. Ehrlich's theory, advanced though it was. For one, he thought the immune cells carried with them sets of keys called "side chains" that could take the right shape and fit into a lock. This was not right but was still a remarkable guess given his lack of technology, and his idea gave rise to one of the single most important words in the language of the immune system. Antibody.

For all the wonder of this discovery, and I'll tell you much more, there is a telling problem with the name antibody. It suggests antibodies go against the body—*anti-body*.

Don't take my word for it. Even some historians in the field have written that the language is complex, even counterintuitive. "The word contains a logical flaw," reads an authoritative recount of the history of the word. Even more broadly, one pioneering immunologist laughed knowingly when describing the complex vernacular of the immune system and said, "You've got a glossary problem."

This is consistent with a theme you'll hear over and over again in the development of the science of the immune system. This group of scientists, immunologists, would win no awards for marketing. They wouldn't be allowed anywhere near Madison Avenue with words like *antibody* and *antigen, macrophage, phagocytosis, glial cell,* and on and on.

Dr. Ehrlich also discovered a universe of different cell types, ones with different edges and shapes and seemingly different functions—and broadened the peculiar language of immunology with cell names like basophils and neutrophils.

Were they part of our defense or something else?

Over time, questions and observations piled up. No wonder. The immune system is one of the world's most complex organic systems, equaled perhaps only by the human brain, with its origins long preceding the evolution of our species.

The distant echoes of its beginnings can be found 3.5 billion years ago, roughly when bacteria, the first cellular organisms, appeared. Using sophisticated chemical and molecular tools, scientists have discovered that some bacteria appear to have sophisticated immune systems, which include the ability to identify specific alien threats and encode memories of them so that, upon invasion, these can be neutralized.

Then about 500 million years ago, a split occurred, resulting in what would evolve into two major immune system lineages. One lineage belongs to non-jawed vertebrates, such as the lamprey and the hagfish. They developed a defense network that is both fundamentally different from ours and nearly as sophisticated. By comparison to ours, theirs is like an ancient language with different lettering, an alternate scripting of the genetic code that confers many of the same defense advantages.

Twenty million years later, around 480 million years ago, the other lineage finds its roots. We know this because creatures that lived that long ago, like the shark, rely on this second lineage. And so do human beings. In the most fundamental sense, we share an immune system with sharks and other jawed vertebrates.

The fact that our version of the immune system has been around that long speaks to its power. Evolution doesn't let things slide that long unless they work.

It is an ever-vigilant, omnipresent peacekeeping force in the Festival of Life.

6

The Festival

Picture a festival—a wide-open, take-all-comers bash. This is life inside your body.

Cells swarm inside you, many keeping to their own areas, regions, organs. They are doing the business of survival and it can be an efficient, well-programmed, though busy, affair. Blood pumps; chemicals flow and fluctuate; conditions change with movement, temperature, thought, emotion, age, and illness; and our invisible machine carries out the orders packaged by sturdy genetic code.

Among these billions of cells, the janitors and manual laborers quietly swarm life's festival, swallowing up detritus and helping rebuild and fix scaffolding after the occasional tissue damage or disruption. They are part of the immune system. So are sentinels and spies that mingle among our cells, picking up signals, rubbing molecule to molecule, collecting data as they brush by, a passive but zealous presence. Is new tissue growing that is cancerous? Is an organ damaged? Are cells spitting out chemicals suggestive of stress to some part of the body, lack of sleep, duress?

And the immune system looks for unwelcome invaders.

Has the body been visited by a pathogen, virus, bacterium, or parasite—perhaps an inhaled ne'er-do-well, or one that entered through a cut in the skin or through invisible excrement insufficiently washed in a bathroom, or picked up on the subway and wiped from the back of a hand into the nose? These pathogens, unlike the healthy cells in our own bodies, don't like to stay in a particular area. They are built to cross borders, push into virgin tissue, spread, eat, and replicate.

Once inside, the pathogen mingles with our cells, reproduces, makes a colony. It takes over an edge of the party and spreads. At this point, one or more of a number of first-line immune system cells suspect danger. These have names like neutrophil, natural killer cell, and dendritic cell. They are constituents of a fire brigade. What follows is swelling, pain, fever. This is inflammation. In the festival of your life, a bar fight has broken out—not yet a full-blown war because it is relatively contained, and your immune system aims to keep it that way.

Many different possible scenarios can follow.

By way of example, inflammation intensifies as immune cells show up in force and devour the infection. Some immune cells blow themselves up in the process. Others nip off parts of the infection and carry them away to be assessed in a defense hub called a lymph node. There, the bits of infection are shared with swarms of passing defenders called T cells and B cells. These are the immune system's most advanced fighters; they are, in fact, two of the most effective biological structures in the world. What makes T cells and B cells so remarkable is that they are extremely specific. Each one of the billions of them in your body is tailored through a quirk of genetics to recognize a very specific infection. Once a T cell or B cell finds its evil mate, its infection doppelgänger, it can set in motion a powerful defense, following hard on the innate reaction, bringing defenders trained specifically to bounce out this particular antigen. Explosions! Implosions! Toxic gas attacks! Good guys eating bad guys!

Sounds like good news, right? Not so fast.

Keeping the peace in the Festival of Life is fraught with its own danger. Inflammation is not fun for the person experiencing disease, and it can put us at risk. The immune response can be accompanied by fatigue, fever, chills, and aches and pains. In millions of people, excessive immune response is its own chronic disease. This is why the immune system, all things being equal, is designed foremost to keep the peace. Excessive force ends badly. The skirmish hurts, the festival is interrupted, the party gripped by anxiety. Life's balance has been upset.

It's a nearly impossible line the immune system must walk, trying not to overreact in the face of pathogens that are also honed by evolution to survive. They are the cunning, violent, and sometimes stupidly brutal festival crashers.

These begin attacking before birth, are nasty, and are everywhere.

7

Festival Crashers

As a newborn, in the birthing ward, you are given an injection. The needle punctures your skin, the very first line of your defense network. The threat didn't even come through the line at the party's velvet rope—not through your mouth or nose. It was sliced in through the roof. The steel invades the tissue. It will likely be clean of bacteria. Regardless, it will cause a localized response, a virtual panic among your cells.

Months later, you might get scratched by the family cat. The cat may carry a microbe. So might the mosquito that landed on your crib and punctured your skin. Mobilization again, within an instant, the most sophisticated defense network in the known world explodes into action.

Or if you are born in a developing country, your mother may give you a sip of water. It will have a parasite in it, a worm. The parasite will descend into your gut. It will settle there and feed.

These are the simplest scenarios. It's possible to imagine endless other circumstances, especially when it comes to a pantheon of bad actors that would make of us their food, their sustenance.

Allow me to introduce you to the villains and the challenges they present. They are highly varied, numbering in the thousands, at least. They take myriad shapes and have their own array of tactics and weapons. When I try to imagine their range, I picture the scene from the original *Star Wars* where Han Solo winds up in a fight with a bounty hunter at a bar known as the Mos Eisley Cantina. Nefarious and odd-looking characters fill the party: wind-instrument-blowing band members that look like their bulbous brains are on the outside; a gorilla-resembling alien with cone-like horns; a bounty hunter with a prickly green head; and so forth. They are serial killers and suicide bombers— Ebola viruses, staphylococci, bird flu, pneumonia viruses or bacteria, syphilis spirochetes, smallpox viruses, polioviruses, and on and on.

As a group, they are known as pathogens, agents that cause disease. It is tempting to think of viruses and bacteria as pathogens, and some of them are, but hardly all. Billions of bacteria cells live inside our bodies without causing harm. In fact, the estimates I've seen indicate that as few as 1 percent are likely to make you sick. And there's a very good chance that you have cancer inside you at this moment, but it is essentially harmless. Like any good story, it can be tough to tell good from evil and indifferent.

The dangerous ones, though, would, unchecked, take no prisoners.

First, bacteria. These are likely one of the earliest life-forms, dating to 3.5 billion years ago. What made them early survivors is that they can grow by themselves as long as they have a food source. They are in this way a self-contained unit. They are small. You can fit several thousand bacteria inside a human cell. For such little things, they can be not just deadly but so lethal that they can change the trajectory of human history, shape culture, rewrite the times. The Black Plague, in the fourteenth century, killed 30 percent or more of Europe's population. Black or bubonic plague is caused by one of the deadliest pathogens known to man, *Yersinia pestis*, a flea-borne bacterium named for the man who discovered it in 1894, Alexandre Yersin. Just goes to show, you should be careful what you discover. Here are a few other bacteria you don't want feeding off you: *E. coli*, salmonella, tetanus bacillus, staphylococci, and syphilis spirochetes.

Next up: viruses.

Bacteria, small as they are, though, dwarf viruses. You can fit several thousand viruses in a bacterium.

Some of the nastier viruses are flu, Ebola, rabies, smallpox. A challenge for viruses is that they tend to be able to reproduce and grow only after they have first invaded a cell and taken over the machinery that it uses to replicate itself.

There is a theory about the origin of viruses that helps explain their nature. Perhaps bacteria came first, and then more complex cells. Then, bit by bit, some bacteria shed parts of their genetic material through random mutation and evolution, and some of those less complex organisms found a way to infect and live off cells, including mammal cells. Those viruses survived. A second theory suggests that viruses peeled off and evolved from our cells, excreta from the human self that found a way to live off and inside of us.

Arguably, the most famous virus of our time is the human im-munodeficiency virus, or HIV. It belongs to a special category called retroviruses. These organisms have the ability to invade a cell and then integrate themselves into our DNA. They mix with us. Imagine how vexing that is for the immune system, trying to discern alien from self. Meanwhile, there's another twist: About 8 percent of our genetic material was formed from retroviruses. That means we've mingled with these viruses and they've become part of us, to the point that they can be not only helpful but essential. An example is the placenta, which may have evolved from a retrovirus in such a way that it helped enable the transmission and sharing of material between mother and child.

Finally: parasites.

Parasites can be much more sophisticated than even bacteria, especially the bigger of these noxious organisms.

They are known as eukaryotic, or "protest," parasites, which is the fancy term for organisms that aren't quite evolved enough to be plants or animals. Some are worms. "Tiny slivers in the tree of life," as Eric Delwart, a molecular virologist at the University of California at San Francisco, described them to me.

They sometimes are deadly, like malarial sporozoan parasites, sleeping sickness trypanosomes, and that giant risk in unsanitary conditions, giardia. And parasites are sometimes so deadly that, like the Black Plague, they have shaped human history through their genocidal capacities. Such is the case with malaria, a parasite that divides quickly in the blood, essentially overtaking a circulatory system.

Bacterium, virus, parasite.

These festival crashers share some important commonalities.

The dumbest ones are so eager to reproduce and to use our bodies to feed on or replicate that they wind up killing us—in effect, killing the host. Ideally, from their perspective, they'd infect us and then they'd have us share them with another person and keep jumping from human to human. But if they fail to do that, they just reproduce themselves, without an off switch, until we're toast and they are too. "They're stupid in that they can get carried away and kill all of us," one immunologist told me.

Another commonality is their mobility. They move around and through barriers in our bodies more easily than other cells. In fact, many cells are quite content to stay in their region or organ, their area of the Festival of Life. Pathogens break through the barriers. Bacteria, for instance, can have little tails, called flagella, little motors that give them bursts of acceleration. A salmonella bacterium, for instance, swallowed with food, might use this propulsive tail to burst through the lining of the gut and into the body. It is built to invade.

The next challenge, and it's a big one, is that these organisms are highly *variable*.

Bacteria and viruses replicate very quickly—bacteria can multiply every twenty or thirty minutes, some viruses faster. Each act of reproduction creates an opportunity for a change, a mutation, a moving around of genetic sequences that can turn a virus or bacteria that our body has figured out how to fight into a virus or bacteria that our body does not know how to defend itself against.

The human reproductive cycle gives rise to a new generation roughly every twenty years. We can't possibly survive an arms race with organisms that change at so much more rapid a pace.

Another way to think about it is that bacteria can divide so quickly that if left unchecked, they could take over our entire body in four days. But our own cells divide relatively slowly, such that they create about sixteen new ones from each cell on a given day. Math is working against us.

So how could it be that a single human body could be prepared to deal with so many threats, including ones that *might not even exist yet*? Think of it: Our immune systems need to cope with rapid-fire mutations from reproducing pathogens—or a protein-based life-form from outer space.

This conundrum is amplified by more simple math. We have a limited number of genes. In the 1970s, that number was thought to be around 100,000 genes in the human genome. Since then, we have learned the number is actually much smaller, perhaps 19,000 to 20,000.

How can we possibly defend ourselves?

"God had two options," Jason's cancer doctor told me. "He could turn us into ten-foot-tall pimples, or he could give us the power to fight 10 to the 12th power different pathogens." That's a trillion potential bad actors.

Why pimples? Pimples are filled with white blood cells, which are rich with immune system cells (I'll elaborate in a bit). In short, you could be a gigantic immune system and nothing else, or you could have some kind of secret power that allowed you to have all the other attributes of a human being—brain, heart, organs, limbs—and still somehow magically be able to fight infinite pathogens.

"This is what makes the immune system so profound," Jason's doctor said.

Much of what I'll explicate in this book is that magic, the way we can survive without being just one big pimple.

Meanwhile, though, there are several other fundamental challenges to our immune system—along with the variety and mutability of bad actors.

One such hurdle has to do with the heart. It's a liability. The trouble with such a powerful central circulatory system is that it pumps blood around our entire body, and fast. Blood moves from head to toe in seconds. So if a pathogen gets into the blood-stream, *Whoosh!* This can quickly become a condition called sepsis—infection in the blood—which can be deadly. A major role of the immune system is to keep infection out of our circulatory system.

Another basic structural complication for the immune system comes from the reality of defending a living creature that must have the ability to grow and heal. The body has to regenerate tissue, all of the time, and replace damaged or outdated cells. Take, for instance, the simple example from the birthing ward: When the vaccination pierces the baby's skin, the body must be able to replace that divot of skin. This is the case too when a splinter stabs or the cat bites. Otherwise, we'd just degrade, erode, bit by bit, like a hill of sand in the rain.

In order to heal, our cells must divide, proliferate. This might sound obvious and simple. But it's precarious for the immune system. That's because it must simultaneously allow new tissue to be developed while also watching with enormous care for bad cells, mutations that are rotten, incomplete, or faulty. That's called cancer.

Only in recent years have we learned that the immune system helps with cell division, promoting healing and rebuilding tissue. But in the process of helping rebuild the body, the immune system can have a hard time discerning bad or mutated cells, ones that look much like us, which are mostly self, but part alien. If it can't tell the difference or gets tricked in some other way by the cancer so that it ignores the usual signals that halt the division of malignant cells, what follows is uncontrolled and reckless growth that is disruptive to normal tissue architecture and function. The immune system can wind up protecting the malignancy.

The line that the immune system must walk is a tightrope over an abyss, with death to the left and the right.

Survival depends on knowing what is self and what is alien. The immune system must cope with three major challenges: the variability of bad actors, the central circulatory system that sends rivers of blood throughout our body in seconds, and the need to heal.

And the immune system must do all that without so overheating that it kills us in the process. It walks the most delicate path. It succeeds with the help of peacekeepers so effective that their work could be mistaken for magic.

The last seventy years in immunology have been a pursuit to understand how the trick works, how our defense apparatus does what it does, at the core. This astounding journey took an arc that moved from a crude conceptual understanding of the immune system and worked down to the molecular level. As a result, medicine can now get in on the magic and begin to meddle with your health inside the machine of your elegant defense.

To explain how this all this applies to your health—and that of Jason, Linda, Merredith, and Bob—I will spend the next hundred or so pages telling you the story of scientific discovery. It goes like this, in brief: scientists got an idea about these things called T cells and B cells, started applying big conceptual knowledge through life-saving vaccines and transplants, and then these imaginative and innovative immunologists delved into the tiny fragments of the immune system, the cogs, and built a blueprint of the machine. They understood, as I'll describe, what inflammation is about, and the molecules that make up our communications network. With each advance of science came another practical step, like building medicines by replicating our defense cells, and then would come another extraordinary scientific leap, like the discovery only a few years ago of a second immune system.

You can think of the immunologists as explorers or Argonauts, pick your metaphor. The deeper and further they got beyond the shore and surface, past the conceptual and theoretical and into the detail, the healthier we got, the longer we lived. Their discoveries saved hundreds of millions of lives, and they are impacting your life and health right now.

So join me on a tour of crucial discoveries and their meaning, starting in a shed in England.

8

The Mystery Organ

In 1941, the world was at war and so were the insides of Jacqueline Miller. A slender and beautiful seventeen-year-old brunette, she coughed until her throat turned raw. She carried with her a spittoon to gather the bloody sputum she rejected from her tattered lungs. For four years, she'd battled tuberculosis, and things were growing dire.

She was helped little by her family's relative wealth and the opulent setting in which she lived. Her father, the manager of a Franco-Chinese bank, had secured a placement in Shanghai, helping the family escape from Europe after Nazi Germany's invasion of France. They took a hurried car ride to Italy and lucked onto the last passenger boat out of Trieste. In China, the family lived in a modern cylindrical five-story house with twenty-four servants—"like kings," recalled Jacques Miller, Jacqueline's little brother, who had himself been born in France. He was ten years old, and would go on to make profound discoveries about the immune system.

In the months before Christmas of 1941, Jacqueline's cough got worse. Jacques watched and listened and tried to make sense of it all. "I overheard the doctor speaking to my mother and telling her we know nothing about how infectious diseases are gotten rid of by the body," Jacques told me. Now in his late eighties, his powerful brain has hardly been slowed by time.

Back then, he recalled, a question nagged at him as he watched his sister. "My sister and I lived in the same room, in the same house with Jacqueline. We never got sick. Why was that?"

Tuberculosis is caused by a bacteria characterized by the waxy surface of its cells. It typically invades the lungs and is infectious, but Jacqueline's little brother and sister didn't contract it. Were their bodies not introduced to it, had they fought it off, or were their genetics different such that they weren't susceptible to it in the first place? Why was it that an alien life-form was taking over this little girl's body, growing inside her, as her defenses lay as ravaged and ineffectual as the Polish and French armies?

All good questions that would be answered with time, but the most pressing one was whether anything could be done for Jacqueline.

What they had already tried is almost laughably, painfully primitive. Prior to the war and their move to China, the family spent time in Switzerland, a hub of tuberculosis treatment. The Swiss treated the disease by injecting air into the chest to cause a lung to collapse. The hope was that this would crush the bacteria and then give the lung a period of rest so that it could reset. Later, while the family was in Shanghai, Jacqueline's father would take her for rides in the countryside so she could breathe fresh air. Meanwhile, as her father tried in vain to help her, he also fought in his modest way to battle fascism, secretly helping to smuggle Frenchmen from the French concession onto boats leaving China for Britain.

When Jacqueline took a drastic turn for the worse that December, "she lost a lot of weight. She looked like a skeleton, like a cadaver," Jacques said, looking back. "I felt horrible."

Jacqueline died on Christmas Day.

Three years later, in New Jersey, researchers isolated streptomycin, the first antibiotic that could kill tuberculosis. Selman Abraham Waksman, the head of the lab at Rutgers University where the discovery was made, won the Nobel Prize for its discovery in 1952.

"If only my sister had hung on two more years, she would've been cured," Jacques said.

Indeed, Jacqueline's death came at an inflection point for medicine and immunology. Science was beginning to put disease on the run. It is remarkable now to look back at what once killed us and see the precipice of discovery.

In 1900, for instance, the leading causes of death per 100,000 patients were pneumonia and flu, followed by tuberculosis and gastrointestinal infection. Heart disease and cancer were well down the list. A century earlier, the first publication of *The New England Journal of Medicine* in the early 1800s lists a study of causes of death that includes 942 patients, nearly a third of whom died from consumption. Almost 50 deaths were stillborns, slightly fewer succumbed to typhus, only 5 had cancer, and a single patient, who, well, medicine could do little about, was struck by lightning.

Approximately 60 million people died in World War II: 15 million on the battlefield, while civilian casualties made up the lion's share of the deaths, according to the National WWII Museum. That was about 3 percent of the 1940 global population.

We died and we were killing each other, and science and society wrestled with these problems, but immunology, to this point, was not a big part of the conversation. It was a backwater. The immunologists had lots of hypotheses about how our bodies defended themselves, but our internal systems were largely invisible, given the relatively primitive nature of our technology. The field was poised for an explosion of learning.

Jacques Miller graduated from medical school in 1956, and he was accepted as a research fellow at the Chester Beatty Research Institute in South Kensington, in London. It was an institution, and an era, in which many researchers focused on cancer, in part because more people were dying from it, as they outlived the infections that had killed people for millennia.

There was another reason to study cancer. The atomic bombings of Hiroshima and Nagasaki led to soaring incidences of leukemia. The radiation from the bomb caused cells to change at a ghastly rate, and it damaged DNA so that the new cells were mutated ones. The more the cells changed, the more they turned into enduring cancers, the kinds that prove so elusive to the immune system. In these bombing victims scientists had a pool for experimentation, and they dove in to understand this pitiable new demographic. The focus on cancer wasn't limited to Japan. The atomic explosion catalyzed such research worldwide.

Dr. Miller's discovery of the T cell's origin owes indirectly to the scientific consequences from this research done into radiation and leukemia in mice.

Mice, mice, mice. This bears repeating because the blossoming of this pivotal period of immunology took place in animal research subjects, largely mice. Immunologists, virologists, and others did their work with these plentiful rodents. In the case of leukemia, researchers irradiated lots of mice to give them cancer. They studied which ones contracted cancer and under what circumstances. The idea was to practice on mice to see if there was anything that might be done to help those wretched souls in Nagasaki and Hiroshima who had been so terribly irradiated.

The research at the time also led to what appeared to be an

unrelated curiosity: A small subset of the mice were observed to spontaneously contract leukemia—whether or not they were irradiated. Scientists noted that this spontaneous occurrence of cancer originated in a small, leaflike organ called the thymus.

The name derives from the word *thymos*, meaning "warty excrescence," which in even plainer language is basically a swelling or a node, an outgrowth. The thymus has two sides, vaguely shaped like leaves or butterfly wings, and is located above the breastbone.

The thymus was long thought to be worthless. Utterly, completely without value to human life, a waste of space, a mysterious vestige of evolution, or God failing to fully clean up after creation.

What happened next is quintessential immunology, a combination of accident, brilliantly conceived experimentation, and controversy.

The satellite office to which Dr. Jacques Miller had been assigned outside London in the late 1950s might hardly be called a laboratory. He worked in a shed, no bigger than a one-car garage. The mice he used were kept in cages in a horse stable.

Dr. Miller's first experiment entailed trying to replicate a previous experiment that had found a new strain of leukemia. It came from extracting leukemic tissue from a mouse with cancer, grinding the tissue down, turning it into a liquid, and injecting it into one of the mice that seemed otherwise to have a low propensity of getting leukemia. The cancer, like a virus, spread to the new mouse.

There was a twist. This worked only if the leukemic filtrate was injected into a newborn mouse, not an adult. Why did just newborn mice get sick? Dr. Miller had an idea of how to answer the question.

"I did something nobody else did," Dr. Miller recalled.

Dr. Miller became expert in removing a mouse thymus, performing thymectomies. He wasn't the first, but he took it to extreme lengths, trying all kinds of permutations. In a significant one, he took a mouse and gave it a leukemic filtrate at birth. After a short period of time, he took out the mouse's mature thymus and replaced it with the thymus of a baby mouse. The mouse would promptly get leukemia. In fact, the adult mice contracted cancer at whatever point they got the immature thymus. "If I replaced it at one month after adult thymectomy, two months, three months, six months. One after the other," Dr. Miller said.

This was at least an oddity and interesting, but was it extraordinary? Did this mean the thymus played a bigger role in health than anyone had ever shown?

Dr. Miller stumbled by accident on a revelation. Remember that he had removed the thymus glands from day-old mice to put them into adult mice. Now he had a group of mice with their thymuses removed. They were the supposed throwaways, rodents sacrificed to science. But Miller noticed they weren't just dying; first they were getting terribly, unusually sick. They were losing weight, shrinking—dying wretched, disease-racked deaths. That seemed odd. "When that happens, you want to open them up and see what the hell is going on," Miller said. He discovered lesions spread across their livers. It looked like hepatitis. They had been overtaken by infection.

So now he had two powerful data points. Mice with an immature thymus could get leukemia. Mice with no thymus whatsoever appeared to be defenseless against disease.

Dr. Miller hypothesized the heretical—that the thymus was of tremendous significance. He then took one more key step to prove it. It's a brilliant idea that he nonetheless dismisses as obvious. He took two mice. He removed the thymus from one at birth. Then he took skin from the other mouse and grafted it onto the one without a thymus.

Dr. Miller did this because it had long been known that skin grafts usually failed, because a healthy immune system rejected the foreign tissue. It was not "self." So he presumed that a mouse without an immune system couldn't recognize foreign skin grafted onto its body. Without an immune system, it wouldn't attack the grafted skin.

Dr. Miller hoped that by putting foreign skin grafts onto the baby mice without a thymus, he could prove the link between the immune system and the thymus, an organ that had previously been considered worthless. Here is what he wrote later about his experiment:

"The results were incredibly spectacular. The mice failed to reject such skin," he said. "The grafts grew luxuriant tufts of hair, and, to convince myself, I even transplanted some mice with four grafts, each from a different strain with a different color." He added, "None of the grafts were rejected, and the recipients looked like they had patchwork quilts on their backs."

He ran the mice through a battery of blood tests, sort of like the ones you might undergo if you go to a doctor and get a complete blood count, but much more primitive. Baby mice, deprived of the thymus, had many fewer of the white blood cells with only one nucleus. These already had the name *lymphocytes*.

This must mean, Dr. Miller thought, that these cells had come from the thymus. "Thymus-derived cells," he called them.

Thymus. T. T cells.

Now, more than fifty years later, Dr. Miller still exhibited a great excitement when he shared the story. I could hear his sense of wonder, followed by an undercurrent of pride and frustration as he explained what happened next. The scientific community didn't believe him. At a meeting of the British Society for Immunology in 1961, he showed slides of his patchwork-quilt mice. His findings were dismissed on a variety of grounds: He'd used a bad strain of mice; disease from the horse stables had infected the mice somehow and perverted the results; whatever he'd learned in mice would have nothing to do with humans.

Dr. Miller published a short paper in the prestigious journal *The Lancet*—with the "bold postulate that the thymus was the site responsible for the development of immunologically competent small lymphocytes." It was his Madame Curie moment. This little leaflike organ, considered a waste of space, long since bypassed through evolution, was central to the immune system.

That revelation was huge but partial—because Dr. Miller didn't know exactly what the thymus was doing. That would come later. But now you know a first puzzle piece in the modern era of immunology, along with a crucial piece of trivia about the origin of the T cell and the fact that it is central to survival.

Dr. Miller thought he'd figured out the main player in the immune system. "I thought it was the only cell," he said of the T cell, "and it could do anything."

He couldn't have been more wrong on that count, but not many people were paying attention. While the insular world of immunology included geniuses celebrated among their peers and feted with Nobel prizes, immunologists were not widely acclaimed; for the most part, immunology was not a draw for most serious scientists, not where the action was. For med students, it was a subject to overlook—"one or two pages in a med school textbook," said Dr. Anthony Fauci, director of the National Institute of Allergy and Infectious Diseases at the National Institutes of Health. "It wasn't ready for incorporation into the main body of science."

The immunologists were just getting started.

9

The B-Word

Turn back the clock a decade to 1951, when an eight-year-old boy with an unusual and disturbing medical history showed up at Walter Reed General Hospital in Bethesda. In the prior eighteen months, the boy had suffered at least eighteen bouts of pneumonia and other life-threatening infections. While the boy could fight off some infection—he was still alive, after all—his body seemed largely unable to mount an immune defense.

The doctor who saw him at Walter Reed was an eventual immune system luminary named Colonel Ogden Bruton. Dr. Bruton ran a test to look for antibodies. At the time, there was a broad conceptual understanding that antibodies were involved in the recognition and targeting of infection. Antibodies, to repeat, are keys that help detect and connect to parts of disease. Cells with antibodies circulate your Festival of Life, looking for their malicious matches. These mechanics, and others, were not yet understood at the time the sick boy came into Walter Reed. But the concept of antibodies had been established. Bruton ran a then-cutting-edge test to look for antibodies. Compared to other parts of the blood, antibodies have a relatively weak electrical charge. So the test involved putting blood into an electrical field and separating out a subset of fluid known as gamma globulins, which contain the antibodies.

The eight-year-old boy didn't have any gamma globulins. He

wasn't making antibodies. This was the first known case of primary immunodeficiency. "Its discovery," notes a biography of Bruton published by the U.S. National Library of Medicine, was "likened in importance to the discovery of yellow fever . . . as an epoch-making contribution to medicine." What the boy and the test told the researchers was that when antibodies weren't present, something terrible could happen.

But there was more that made the boy's case vexing. He didn't have antibodies, *but* he still had white blood cells and was still able to fight off some viruses. The boy's thymus was intact too.

This conundrum vexed scientists. What were the defense's main components?

A nasty divide erupted among immunologists about the core source of the body's defenses. One camp thought the antibody was the center of the action. This was a substance, a process, a chemical reaction of some kind that helped attack alien threats. It was called *antibody-mediated immunity.* But others thought the T cell was the center of all the action. Their philosophy was called *cell-mediated immunity.* It meant that these T cells ruled the day.

The centuries-old mystery chicken from Fabricius helped resolve the debate.

In 1952, the year after the boy showed up at Walter Reed, a young scientist at Ohio State University was watching his professor dissect and autopsy a goose. The scientist later wrote that he watched his professor remove the bursa and asked, "What is that? What is its function?"

"Good question. You find out," the professor replied. The scientist noted that with this suggestion, "the search began."

He deduced that the bursa of Fabricius—that seemingly vestigial organ in the back of the bird—grew very quickly in the chick's first three weeks of life. Two years later, in 1954, a fellow researcher discovered that chickens with the bursa removed could not generate a response to vaccine because they made a very low volume of antibodies.

No bursa, few antibodies.

That sure doesn't sound like a vestigial organ either. It suggested that, in birds at least, antibodies might come from the bursa. But humans have no bursa.

The curiosity would be resolved in part by Dr. Max Cooper, a physician shaped, like Dr. Miller, by painful historical reality. His biography isn't a sideshow. It's part of the immune system story.

Dr. Cooper grew up in the 1940s and early 1950s in rural Mississippi. He lived in a tiny town in which he worked at every kind of odd job—as a janitor at the school, behind the counter at the drug store, in the oil field, delivering newspapers. His parents were unusual in that they had high levels of education and so, to young Max Cooper, the most revered man in town was the doctor—"the pinnacle of society," he recalled. Max knew what he wanted to do.

He graduated from medical school at Tulane, where, during his final year, he saw a patient with digestive problems. The man was a conductor on the *Panama Limited,* a train that traveled between Chicago and New Orleans. He was distinguished.

And he was on the "colored" side of Charity Hospital in New Orleans, because in those days the hospital was segregated. Dr. Cooper examined the patient and then made a presentation to the senior doctor, called the attending physician.

"Mr. Brown's chief complaint is that—" he began before the attending physician interrupted.

"Who told you to call this nigger *Mr.* Brown?" the physician said. "Would your father have taught you to call this nigger *Mr.* Brown? We don't do that here at Tulane."

"Yes, sir," Dr. Cooper responded, and then spent a lifetime regretting he had not responded differently.

In 1960, whites in the United States lived about 70.5 years on average. Nonwhites, which was the other broad category measured by the government, lived on average to 63.5. There were lots of contributing factors, including environment and its interaction with the immune system. Scientific revelations about this would come later. Also worth noting at that time, women lived longer (75 years) than men (66.5 years), a disparity consistent in whites and nonwhites.

Dr. Cooper began to think about the differences among people, and their defenses. And as you'll see, culture, environment, discrimination, all of it contribute to individual and societal identities, how we define our communities, see self, and nonself, ideas that are core to how the immune system polices our bodies but also how we define and police our societies.

By now it was the mid-1960s and Jacques Miller had published his seminal work about the thymus. At the University of Minnesota, Dr. Cooper, fascinated by the emerging debate about the immune system, became interested in a rare disorder you'd wish on no one. It is called Wiskott-Aldrich syndrome. The patients suffer severe immune deficiency.

"They could get a fever blister, and if their body couldn't control it, it became a widespread infection that killed them," Dr. Cooper said. They typically died within three years.

Cooper started to study the autopsy reports. Again, he found this conundrum: There were plenty of white blood cells—lymphocytes—but very few antibodies. The thymus seemed to be working, but for the most part, the overall immune system was *not* working.

That's when it hit him. "There were two lineages of lymphocytes," he said. In other words, the T cell wasn't the only game in town. The immune system wasn't connected only to the thymus. There must be more.

One clue had come from the chicken. Without a bursa, the chicken had many fewer antibodies. To hone in on the answer, Dr. Cooper and his colleagues experimented on chickens and discovered that indeed, one set of immune cells appeared to come from a chicken's bursa and another from the thymus. So now the two parts of a chicken's body that had seemed to have had no purpose were now seen as key to producing a lineage of immune cells.

But humans aren't chickens (thank you, author!). We have no bursa. So where might *our* antibodies come from?

A next clue came from researchers in Denver who were experimenting with (what else?) mice. They discovered that even when a mouse lost its thymus, it could still mount some defense. And the defense appeared to originate from the bone marrow in the mouse.

One of the researchers theorized that the cells from the thymus and the cells from the bone marrow were working together. Perhaps, the researcher thought, cells from the thymus could somehow produce the antibody but only with help from the cells originating in the bone marrow.

The researcher added: "These are not problems which the present analysis can resolve."

Jacques Miller was back on the case. He helped put the final pieces together.

"It's very complicated to describe," Dr. Miller told me by phone from Australia. "It will be hard for you to understand."

"Try me."

"It's a very, very classic experiment."

He attempted to describe his seminal experiment linking T cells and B cells. He tried me. I will not try you. It is indeed extremely complicated, involving the creation of a hybridized mouse of two different strains—mixing and matching bone marrow and thymus, and looking for the source of immune system cells.

What Dr. Miller found out "changed the course of immunology!" he wrote to me in an email, and he wasn't bragging. It was true. (And it is also true that there were many other crucial contributions to the subject made by other scientists at the time.)

Miller's complex experiment helped show that one set of immune system cells came from the thymus and another from the bone marrow. There were differences between these types of cells that defined the relationship between them. The T cells began in the bone marrow and then moved to the thymus, where they matured. They seemed to be very authoritative cells. The T cells could fight disease or infection directly.

Then there are the B cells. They originate in the bone marrow. These cells were what Dr. Miller called "antibody-forming precursor cells"—they were ready to be armed in some way to fight disease. But it appeared that B cells required some instruction, some additional information to act. That information seemed to come from the T cells, which were instructing other cells in how to attack.

The B cells came from bone marrow and generated antibodies. The T cells matured in the thymus and could either fight or direct action. They are generals and soldiers.

At least that was the theory at the time. There was a lot of validity to it, as well as even more missing information.

Dr. Miller strove to generate clever names for these two lineages of immune fighters. He couldn't come up with anything particularly clever or useful. Several years later, though, they got their names from a connection that seems obvious to us now. The B cells come from the *b*ursa or *b*one marrow, and the T cells from the *t*hymus, and "since then, hardly an article has appeared in any immunological journal without mentioning the words T cells or B cells," Dr. Miller would later write.

This was wonderful and also theoretical. A T cell, a B cell. Nifty names. How did they function? If they worked together, how did they communicate?

10

T Cells and B Cells

Now you know how the T cell and B cell got their names. Still, the breadth of their purpose would take decades to understand, with nuance added virtually every year. For a long time, conceptually, the T cell and B cell were considered the core of the immune system and, to some, its only part.

It turns out they are both essential yet also heavily reliant on another group of potent killer cells, as well as an array of communications and surveillance systems.

But what *are* the T cells and B cells, and what's this got to do with *you*?

Remember the milky-white veins discovered by Gaspare Aselli during his dissection of the dog in 1622? The white substance is made up of white blood cells. Some of these are T cells and some are B cells—with other cells in the mix too.

Broadly, white blood cells are different in key ways from the red blood cells that most of us associate with "blood." Red blood cells, for one thing, appear red, not white. So there's that. The two kinds of cells also have fundamentally different contours. Red blood cells look like beautiful circles carved with graceful indentations. White blood cells resemble baseballs covered in spikes. Many of these spikes are receptors. They send and receive signals. These cells are information hubs, and they can be vicious killers.

White blood cells are essential for your survival. They are as vital to life as the red blood that carries oxygen. T cells and B cells are the most specialized part of the system. They are particularly crucial when you face a complex or unusual bacteria or virus. This is because these B cells and T cells are incredibly targeted. They are the cells able to manufacture precise killers tailored to specific diseases. Within your sea of white blood cells is a match for virtually any pathogen that infects you, and a big key to your health involves the speed with which the right T cell and B cell can make contact with the disease, bind to it, and then manufacturer tens of thousands of copies of the precise defender to wipe out the offenders.

Let's say it's flu season. You're on an airplane or a bus, and someone coughs. You're in your cubicle at work. You're a full five feet away from the infected person. Not far enough, says the Centers for Disease Control and Prevention (CDC), which puts the flu's range of travel by sneeze or cough at six feet. Or you can get flu on your skin through a touch on a handrail that a carrier has touched not long before. A kiss, a hug, a handshake. You wipe your nose, and now the virus has a warm and comfy place to reproduce.

Almost immediately, the immune system picks up an intruder, but at this point in the scientific journey—in the chronology of discovery—immunology didn't really understand what first contact looked like. That came later.

So, back to the flu and you, and T cells and B cells. When you are first infected, your body generates a kind of generic response. It is during this period that your elegant defense is waiting for your T cells and B cells to generate a powerful response. The delay can take five to seven days. That is because the right B cell and T cell, with the right antibody or receptor, must be contacted, or make contact with the bug, fit lock into key, and begin generating defenders. Many times, then, the best case is that you're sick for a

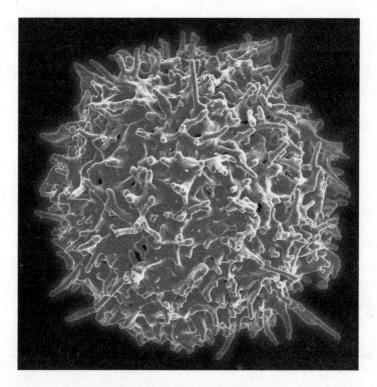

A T cell, central to our elegant defense. (NIAID/NIH)

few days while this immune response kicks in. Again, this doesn't mean you're without defenses until that point, but it means you're without precision defenses, like a T cell or a B cell.

What we know now is that T cells and B cells find their prey in very distinctive ways, and those distinctions themselves are crucial to understanding the complex evolution of the immune system.

On the surface of T cells, some of the spikes are able to identify the signature, or fingerprint, of pathogens, the bad guys. However, for the most part the T cells don't recognize the pathogen directly. They do so through an intermediary that I'll introduce shortly in a fuller context. For now, suffice it to say, the T cells get a message

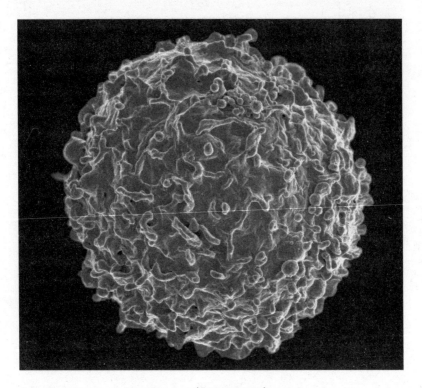

A B cell, originating in bone marrow. (NIAID/NIH)

alerting them of the presence of a dangerous intruder. When that happens, T cells can take on different roles. Some are foot soldiers and others are generals. The generals can dispatch other T cells to the front lines. Or they can send B cells into battle.

B cells can also recognize pathogens more directly using a special kind of receptor called an antibody. Antibodies are protein molecules with extraordinary abilities, and they are central to the immune system.

Antibodies sit on the surface of B cells. They help identify pathogens by acting like a combination of an antenna and a house key.

Like an antenna, antibodies pick up signals. But each antibody is finely tuned. It picks up only one type of signal. In fact, so particular is each antibody that most of the billions of white cells coursing through us generally have unique antibodies on their surfaces. So unlike most antennae—say, radio towers—the antibody receptor doesn't pick up just any signal. It picks up one. It is evolved to connect to a single kind of organism.

The antibodies on the surface of these cells discover the organism that is their match, or mate, by running into it. Literally crashing into or rubbing onto it. These white cells course through the body, through the raucous festival within us, and they roam and flow and jumble and can spend years of restless irrelevancy until one day—*Boom!*—they smack into the chemical structure that they, and only they, can attach to.

What the antibody attaches to is its own little nub or receptor on a cell. The thing it attaches to is called an antigen. An antigen is the mate to an antibody. The antibody and the antigen bind to each other, like a lock and a key.

If you get a bacterial infection, the pathogen trying to spread through your body expresses a particular antigen. Inside your body, there is a B cell that discovers the antigen, binds to it, and annihilates it. Or it sets off a cascade of other defenses.

Even before science knew all these things, one absolutely essential and common trait of the T cell and B cell stood out: they can learn. These cells are highly adaptive, which is why they are referred to as the "adaptive immune system."

This capability to adapt explains a practical development that is one of the most important life-saving discoveries in our species's history. Enter the vaccine.

11

Vaccines

Vaccines are a boot camp for the immune system. The inoculations prime and teach the immune system, effectively training the T cells and B cells and giving them a cheat sheet. The right vaccine can provide your body with the power to mount a faster response to diseases that could otherwise be deadly or devastating in other ways, whether smallpox or polio.

It's not that our elegant defenses won't mount an attack against these diseases absent a vaccine, but the attack might well be insufficient given the time it takes for the immune system to identify the bug and start manufacturing enough soldiers to fight back. In the meantime, you might well die. That said, it's no small thing to find the right vaccine. The lesson of this chapter is that the immune system can learn, but it's not easy to teach.

Among the most famous names in vaccines is Edward Jenner, the English doctor who developed the smallpox vaccine. Less well known is that the groundwork for Dr. Jenner's discovery had been laid through various experiments aimed at stopping smallpox, the variola virus, which appears, according to the CDC, to have been around since Egyptian times (evidence: mummies with pustule scars).

Smallpox was spread through the air, by sneezes, coughs, or close interaction with a victim. It killed 30 percent of those who contracted it. Its lethality has to do with the way it and related viruses pull a stunt on the immune system. The infections can block the transmission of a distress signal that calls killer immune cells into action. (I'll save this discussion of the way diseases trick the immune system because it relates in no small way to how immunology helped save Jason.)

Prior to the work done by Dr. Jenner, the effort to control smallpox was called variolation, the name drawn from the name of the virus. If you think vaccinations seem unpleasant nowadays, this precursor was worse. "Material from smallpox sores (pustules) was given to people who had never had smallpox. This was done either by scratching the material into the arm or inhaling it through the nose," the CDC notes in a history of the technique. Unpleasant though it might've been, it did curb the likelihood of getting the disease in some people, though not enough to stop its epidemic spread.

To physicians and scientists of the period, it showed that the immune system seemed able to develop a response that could later be called into play. The system can acquire a cheat sheet that both helps to quickly identify a problem and has the instructions on how to immediately liquidate the foe. Variolation usually didn't work, though. In most cases, the immune system didn't get sufficiently educated to, or stimulated against, smallpox.

Then came a turning point for medicine.

The setting was Gloucestershire, England, in 1796. It is hallowed ground well-trodden in the history books. Dr. Jenner noticed that the cow's milkmaids had pustules but didn't seem to get the deadly disease. From a cowpox lesion of a milkmaid, he poisoned an eight-year-old boy. The boy lived. Somehow this cowpox strain was the right varietal to spark an immune system defense. Happy birthday, world's first vaccine!

Even then, though, scientists understood that there was a corollary to the immune system's ability to learn: It is not easily taught. As often as not, efforts to create vaccines failed. The concoction seemingly had to be perfect. Little changes could render inoculations ineffective. Researchers discovered that successful vaccines were strong enough to provoke a powerful response by the immune system, but weak enough—*attenuated* is the scientific term—to keep it from being as nasty as the infection itself. The wrong combination entailed the risk that, instead of protecting, a vaccine could kill.

That's what happened with the initial mass test of the polio vaccine.

The first polio epidemic was recorded in 1894, 132 cases in Vermont. Of the infected, 1 to 2 percent were paralyzed.

The poliovirus gets quickly into the bloodstream, after entering through the mouth and growing in the throat and gastrointestinal tract. It winds up in the nervous system, where it attaches to nerve cells and invades them. It then takes over the nerve cell's manufacturing process to reproduce itself—thousands of copies in an hour. Then it kills the cell and moves on to infect others. Picture a shadow creeping over our festival as cell after cell goes dark.

The vexing effort to eradicate polio included the work in the 1930s of two competing scientists, Dr. Maurice Brodie, a Canadian working at New York University, and a Philadelphia pathologist at Temple University in Philadelphia named Dr. John Kolmer. Various histories I'll delineate here recount their failings, even disasters.

The two competing scientists had similar ideas. They infected monkeys with polio and tried to make a human vaccine with the nerve tissue. In Brodie's case, he then mixed the liquefied monkey tissue with formaldehyde, called formalin, hoping to "deactivate" the virus. It would present enough, the theory went, to provoke an immune response, but would not be powerful enough to actually infect. Not so much. One history, written by a Yale doctor and historian named John Paul, is quoted as saying that Brodie's vaccine was tested on 3,000 children, but "something went wrong, and Brodie's vaccine was never used again." A history published in the *New York Times* is more explicit: Children were left paralyzed.

Dr. Kolmer had the same results, though he took a slightly different approach. He took the monkey nerve tissue, mixed it with chemicals, and refrigerated the mixture to attenuate it. Dr. Paul's history calls it a "veritable witch's brew." More infected children. In Paul's book, Dr. Kolmer is reported to have said at a public health conference in 1953: "This is one time I wish the floor would open up and swallow me."

In 1952, as *Time* magazine reported, the worst outbreak yet infected 58,000 Americans, killing 3,000 and paralyzing 21,000. "Parents were haunted by the stories of children stricken suddenly by the telltale cramps and fever," *Time* read. "Public swimming pools were deserted for fear of contagion. And year after year polio delivered thousands of people into hospitals and wheelchairs, or into the nightmarish canisters called iron lungs."

The answer to the polio mystery, also well known, came from Jonas Salk, who was born in New York City of Russian Jewish immigrant parents and eventually was appointed director of the Virus Research Laboratory at the University of Pittsburgh School of Medicine (by way of New York University and Michigan). His vaccine weakened the poliovirus with formaldehyde and mineral water. It effectively "killed" the poliovirus. But it was recognizable enough for the immune system to pick it up. Ta-da! It cut the risk of infection in half.

The country scrambled to produce and disseminate the vaccine as quickly as possible. Alas, this happy ending comes with an asterisk. The first big batch of vaccine wasn't properly made. Cutter Laboratories in California, one of the main producers of the vaccine, inoculated more than 200,000 children in 1955, and within days there were reports of paralysis. Within a month, the program was discontinued, and investigations revealed that the Cutter vaccine had caused 40,000 cases of polio, leaving 200 children with varying degrees of paralysis and killing 10.

These problems were ironed out and polio was all but eradicated in the United States and, eventually, worldwide. Here's the lesson: Intervening on behalf of the immune system is no easy task, given the delicate balance. The vaccines were the first big step in that direction, even if we didn't truly understand their dynamic. Without fully understanding the mechanisms, we had found an effective tool.

That was true of the discovery of a second, marvelous immune system ally: antibiotics.

Antibiotics are arguably more important than vaccines. In fact they are "probably the most successful forms of chemotherapy in the history of medicine. It is not necessary to reiterate here how many lives they have saved and how significantly they have contributed to the control of infectious diseases that were the leading causes of human morbidity and mortality for most of human existence," according to a history published in a National Institutes of Health journal. Broadly, antibiotics work by taking advantage of

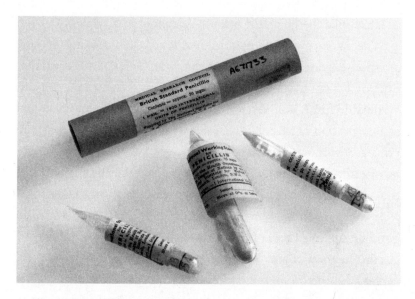

Early doses of penicillin, a medicine that changed the world. (Science Museum, London/Wellcome Collection)

differences between human cells and bacterial cells; for instance, bacterial cells have walls that human cells do not. Antibiotics can prevent bacteria from building such walls.

That's the mechanism behind the shot heard 'round the world in 1928, at St. Mary's Hospital at the University of London. The world was temporarily at peace, which was fine with a Scotsman named Dr. Alexander Fleming. He'd seen plenty of the opposite in the Army Medical Corps during the Great War.

The accident took place in a petri dish. It was filled with the strep bacteria he was studying. One day he noticed something odd. One area of the dish containing the deadly pathogen was suddenly free of bacteria. A closer looked showed it was being killed off by mold—"the mold had created a bacteria-free circle around itself," read Fleming's Nobel Prize biography in 1945. Why the Nobel?

He named the medicine born of that mold *penicillin*.

Whereas vaccines prompt our own response, antibiotics import a response from the outside, and that is an absolutely critical distinction for our everyday health. The reason is that when you add an outside force, you disrupt the natural order. Even if the goal is preservation of life, and even if it works, that doesn't mean the process is without important risks. In the case of antibiotics, these terrific killers don't just kill off bad bacteria; they target good stuff too, including bacteria crucial to your health and well-being.

If you've ever taken antibiotics and gotten diarrhea, you're in good company. The antibiotics are killing off bacteria in your gut that help you digest. They are doing real damage inside your gut, even as they get rid of the pathogens that could turn off the lights in your Festival of Life. Later, I'll get deeper into the importance of the daily and long-term health of your gut, the microbiome, but at the time that antibiotics first emerged and became wonder drugs, the concept was more basic: survive the infection to fight another day.

Now, because of Dr. Fleming, you wouldn't die from getting a cut on your hand, or a small battlefield wound, ear infection, and on and on. Antibiotics have not only extended life but improved its quality by permitting myriad modern surgical procedures, like knee and hip replacements, which would be at extreme risk of infection absent these wonder drugs. Plus, antibiotics are used to keep livestock healthy, helping grow the food supply.

But vaccines and antibiotics weren't easy to come by, at least not effective ones. The body had to do most of the work. It had been doing most of the work—for epochs.

Plus, the immunologists pioneering the exploration were determined to get deeper into the machine, for both intellectual reasons and practical ones—could they figure out how to extend life further and further? That meant answering the biggest question of all: how could our bodies become equipped with defenses for so many possible threats? How is it we can survive in a world in which the possible threats are practically infinite?

12

The Infinity Machine

It's vacation time. You and your family visit a country where you've never been and, in fact, your parents or grandparents had never been. You find yourself hiking beside a beautiful lake. It's a gorgeous day. You dive in. You are not alone. In the water swim parasites, perhaps a parasite called giardia. The invader slips in through your mouth or your urinary tract. This bug is entirely new to you, and there's more. It might be new to everyone you've ever met or come into contact with. The parasite may have evolved in this setting for hundreds of thousands of years so that it's different from any giardia bug you've ever come into contact with before or that thrives in the region where you live.

How can your T cells and B cells react to a pathogen they've never seen, never knew existed, and were never inoculated against, and that you, or your doctors, in all their wisdom, could never have foreseen?

This is the infinity problem.

For years, this was the greatest mystery in immunology.

Of course, the immune system has to neutralize threats without killing the rest of the body. If the immune system could just kill the rest of the body too, the solution to the problem would be easy. Nuke the whole party. That obviously won't work if we are to survive. So the immune system has to be specific to the threat while also leaving most of our organism largely alone.

Over the years, there were a handful of well-intentioned, thoughtful theories, but they strained to account for the inexplicable ability of the body to respond to virtually anything. The theories were complex and suffered from that peculiar side effect of having terrible names—like "side-chain theory" and "template-instructive hypothesis."

This was the background when along came Susumu Tonegawa.

Tonegawa, like Jacques Miller, was born in 1939, in the Japanese port city of Nagoya, and was reared during the war. Lucky for him, his father was moved around in his job, and so Tonegawa grew up in smaller towns. Otherwise, he might've been in Nagoya on May 14, 1944, when the United States sent nearly 550 B-29 bombers to take out key industrial sites there and destroyed huge swaths of the city.

Fifteen years later, in 1959, Tonegawa was a promising student when a professor in Kyoto told him that he should go to the United States because Japan lacked adequate graduate training in molecular biology. A clear, noteworthy phenomenon was taking shape: Immunology and its greatest discoveries were an international affair, discoveries made through cooperation among the world's best brains, national boundaries be damned.

Tonegawa wound up at the University of California at San Diego, at a lab in La Jolla, "the beautiful Southern California town near the Mexican border." There, in multicultural paradise, he received his PhD, studying in the lab of Masaki Hayashi and then moved to the lab of Renato Dulbecco. Dr. Dulbecco was born in Italy, got a medical degree, was recruited to serve in World War II, where he fought the French and then, when Italian fascism collapsed, joined the resistance and fought the Germans. (Eventually, he came to the United States and in 1975 won a Nobel Prize for using molecular biology to show how viruses can lead, in some cases, to tumor creation.)

In 1970, Tonegawa—now armed with a PhD—faced his own immigration conundrum. His visa was set to expire by the end of 1970, and he was forced to leave the country for two years before he could return. He found a job in Switzerland at the Basel Institute for Immunology.

Around this time, new technology had emerged that allowed scientists to isolate different segments of an organism's genetic material. The technology allowed segments to be "cut" and then compared to one another. A truism emerged: If a researcher took one organism's genome and cut precisely the same segment over and over again, the resulting fragment of genetic material would match each time.

This might sound obvious, but it was key to defining the consistency of an organism's genetic structure.

Then Tonegawa found the anomaly.

He was cutting segments of genetic material from within B cells. He began by comparing the segments from immature B cells, meaning, immune system cells that were still developing. When he compared identical segments in these cells, they yielded, predictably, identical fragments of genetic material. That was consistent with all previous knowledge.

But when he compared the segments to identical regions in *mature* B cells, the result was entirely different. This was new, distinct from any other cell or organism that had been studied. The underlying genetic material had changed.

"It was a big revelation," said Ruslan Medzhitov, a Yale scholar. "What he found, and is currently known, is that the antibody-encoding genes are unlike all other normal genes."

The antibody-encoding genes are unlike all other normal genes.

Yes, I used italics. Your immune system's incredible capabilities begin from a remarkable twist of genetics. When your immune system takes shape, it scrambles itself into millions of different combinations, random mixtures and blends. It is a kind of genetic Big Bang that creates inside your body all kinds of defenders aimed at recognizing all kinds of alien life forms.

So when you jump in that lake in a foreign land, filled with alien bugs, your body, astonishingly, well might have a defender that recognizes the creature.

Light the fireworks and send down the streamers!

As Tonegawa explored further, he discovered a pattern that described the differences between immature B cells and mature ones. Each of them shared key genetic material with one major variance: In the immature B cell, that crucial genetic material was mixed in with, and separated by, a whole array of other genetic material.

As the B cell matured into a fully functioning immune system cell, much of the genetic material dropped out. And not just that: In each maturing B cell, *different* material dropped out. What had begun as a vast array of genetic coding sharpened into this particular, even unique, strand of genetic material.

This is complex stuff. But a pep talk: This section is as deep and important as any in describing the wonder of the human body. Dear reader, please soldier on!

Researchers, who, eventually, sought a handy way to define the nature of the genetic change to the material of genes, labeled the key genetic material in an antibody with three initials: V, D, and J.

The letter V stands for variable. The variable part of the genetic material is drawn from hundreds of genes.

D stands for diversity, which is drawn from a pool of dozens of different genes.

And J is drawn from another half dozen genes.

In an immature B cell, the strands of V, D, and J material are in separate groupings, and they are separated by a relatively massive distance. But as the cell matures, a single, random copy of V remains, along with a single each of D and J, and all the other intervening material drops out. As I began to grasp this, it helped me to picture a line of genetic material stretching many miles. Suddenly, three random pieces step forward, and the rest drops away.

The combination of these genetic slices, grouped and condensed into a single cell, creates, by the power of math, trillions of different and virtually unique genetic codes.

Or if you prefer a different metaphor, the body has randomly made hundreds of millions of different keys, or antibodies. Each fits a lock that is located on a pathogen. Many of these antibodies are combined such that they are alien genetic material—at least to us—and their locks will never surface in the human body. Some may not exist in the entire universe. Our bodies have come stocked with keys to the rarest and even unimaginable locks, forms of evil the world has not yet seen, but someday might. In anticipation of threat from the unfathomable, our defenses evolved as infinity machines.

"The discoveries of Tonegawa explain the genetic background allowing the enormous richness of variation among antibodies," the Nobel Prize committee wrote in its award to him years later, in 1987. "Beyond deeper knowledge of the basic structure of the immune system these discoveries will have importance in improving immunological therapy of different kinds, such as, for instance, the enforcement of vaccinations and inhibition of reactions during transplantation. Another area of importance is those diseases where the immune defense of the individual now attacks the body's own tissues, the so-called autoimmune diseases."

These last sentences—about autoimmunity and transplantation—introduced a central challenge to understanding our body's defense: How does something so powerful avoid attacking the healthy parts of us? And how can we treat ourselves medically through things like transplants and medication without risking that our potent antibody system will reject anything that could actually help us, even if it seems alien at first?

What is alien, and what is self?

13

Transplant

One day in the early 1970s, a family showed up at the Mayo Clinic in Rochester, Minnesota, with an infant boy suffering a mysterious condition. The baby had skin lesions that looked like measles, terrible diarrhea, and a fever. The petrified parents had further reason to be afraid. They'd previously had another child, a baby boy ailing just like this one, and he had died in mere weeks. That older boy, after his death, was discovered to have neither B cells nor T cells.

The family brought their second infant to Mayo seeking a consultation with Dr. Max Cooper, the prominent immunologist who had helped explain the presence of T cells and B cells. Could he save their baby?

"We were asked to make a suggestion," Dr. Cooper recalled. He felt a dire sense of responsibility but had little science to go on. What if, he suggested, they withdrew bone marrow from the boy's mother and then injected it into the bone marrow of the boy?

Bone marrow is home to maturing cells, stem cells, including maturing immune system cells. Could it be that the mom's immune system could take hold in her baby boy?

There wasn't much evidence to tell them what would happen. "We had only a crude road map," Dr. Cooper recalled. The transplant needed to succeed in killing the foreign disease without causing the boy's own hobbled defenses to react against the mom's immune cells. Cooper theorized that the boy's body wouldn't reject his mother's defenses because they had so much genetic material in common.

With Cooper monitoring by phone from the nearby University of Minnesota, the mother's cells were pulled from her hip with a long needle and were then injected into the ailing boy. Twelve days later, the boy developed a fever and lesions that looked like measles. The diarrhea returned. The boy died.

For all Dr. Cooper's brilliant work in his career, he was left defenseless and scarred by the immune-challenged boy. The boy couldn't fight off the disease and his condition was perhaps made even worse ultimately by the introduction of the mom's foreign cells.

"The boy was clearly going to die without our doing something, but the idea that I pulled the trigger is a pretty awful feeling," Dr. Cooper recalled.

Why wasn't it possible to just replace one immune system with another? Imagine how elegant and simple it would be to extract the T cells and B cells from a healthy and thriving person and inject them into a person whose body was failing to combat disease.

Or for that matter, how wonderful to be able to transplant healthy skin from one person onto the diseased and gangrenous leg of, say, a felled soldier? Why aren't our parts interchangeable?

I visit the idea of transplantation here for two reasons. One is that it helps explain the challenge of interchanging our parts. The other is that it shows the interplay between the deepening scientific exploration by immunologists and the practical implications of their work. With each successive discovery, not surprisingly, came more life-saving and life-improving procedures and medicines. The two, revelation and real-life use, fed on each other as the twentieth century moved ahead, the Argonauts of the immune

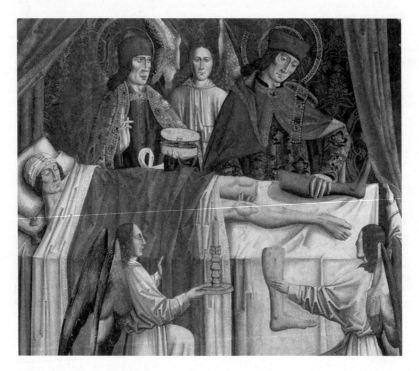

Saint Cosmas and Saint Damian, patron saints of transplantation. Infallible they weren't.

system journeying deeper and deeper, finding treasure and tools and putting them to use. Few examples illustrate this as well as transplantation.

The idea of transplantation has long been a perilously seductive and deadly prospect. The reasons for its ultimate complexity tell us about an essential balancing act necessary for the survival of the species. Humans must be incredibly similar and also fundamentally diverse. The similarity is necessary to allow us to work together, communicate—share resources, ideas, food. But we must be different enough to provide diverse talents, including the innate ability to fight off different threats. Simply put, if all of our defense systems were the same, a single deadly disease could come along and wipe out the lot of us.

This tension between sameness and individuality leads to certain trade-offs. One of them is that we cannot easily exchange our parts—say, my leg for yours or my immune system for yours. In fact, a defense system that works wonders protecting one person can be quite deadly in the life of another.

The broader history of transplantation helped yield key clues to the extraordinary specificity of our defense networks.

That history includes the legend of two Catholic saints, twin brothers, Saint Cosmas and Saint Damian, who were miracle workers in the third century—in their own minds. The saints reported transplanting an entire leg from one person to another—a "successful" early transplant, according to two of the world's current transplant leaders, Dr. Clyde F. Barker at the University of Pennsylvania and Dr. James F. Markmann at the University of Massachusetts, who wrote of this miracle in a rich and wonderful history of transplantation.

They used quotation marks around the word *successful* because the transplanted leg miracle was actually an abject failure. (No, these self-proclaimed miracle-working brothers didn't successfully sew one person's leg onto another one's body, which maybe is why Cosmas and Damian were later made the saints of pharmacy, not saints of transplantation.)

Over the centuries, stories abounded of successful transplants, like skin flaps used to replace missing noses. "Legends and claims of miracles," the two transplant specialists wrote in their history. "Centuries of sloppy observation and self-deception." It sounds carnivalesque, the stuff of hair tonic salesmen.

The reason these attempts didn't work has to do squarely with the immune system.

"Transplant is the parent and sibling of immunology," Dr. Markmann told me when we spoke about the history. Transplantation and immunology are siblings because a transplant—whether of skin or an immune system—can't work if the body rejects the transplant as "alien." And transplantation is also immunology's parent because the failure of the body to accept human or animal tissue was one of the clearest, earliest signs that something in our bodies was rejecting and attacking tissues that seemed so similar. It was a clue to the power and precision of our elegant defense.

Science experimented with all kinds of transplants without success. Human transplants of kidneys were tried and failed; dog kidney transplantation was a flop; and the ethics of the field were questioned.

At last, though, came a scientific breakthrough, thanks to a zoologist.

Peter Medawar—eventually *Sir* Peter Medawar—was a zoologist from Oxford who had been called upon during World War II to help a plastic surgeon treat burn victims. Sir Medawar tried in vain to graft skin from donors onto the charred victims of shelling and bombing. The results were cruel. Even the failed grafts looked for several days or weeks like they were succeeding. That's because skin doesn't have as many blood vessels and as much blood flow as, say, kidneys or other internal organs. It would take time for the immune system cells, carried in the blood, to assess and reject the skin.

"The skin would sit there and look pretty good," Dr. Markmann said; the soldier and Dr. Medawar would feel cautiously optimistic. "Then the graft would turn sour. It looked so nice and happy, and it would ultimately always fail."

Many of these battlefield and deathbed stories would lead me to realize a hard reality about science and scientists, especially immunologists. Often, great discoveries have been made on the cusp of death through experimentation on a patient. The patient would be complicit, usually agreeing to take a chance at living through a Hail Mary. Revelation had the deeply perverse quality of being born of desperation, not just from science's grand yearning to save lives, but from the excruciating intrapersonal despair that allows a human to become a guinea pig. *Experiment on me so that I don't die.* This would eventually become clear to me as I watched Jason on the precipice of death, putting himself in the hands of a well-meaning, informed oncologist who was half blinded by science's limitations.

After World War II, Dr. Medawar continued his transplant work—on rabbits and humans alike. He tried performing skin grafts between siblings. Presumably, immune systems would be less likely to reject tissue with related genetics, like that of siblings. Would you believe, though, it sometimes made matters *worse*? Take the example of a woman burned in a gas stove incident who was brought to a clinic and given a skin graft from a sibling. After a few weeks, the graft was rejected. In some cases, a second sibling graft was tried, yielding a surprise: rejection happened faster, from, say, two weeks down to a week.

In other words, the immune system was more deliberate the

first time around in rejecting a sibling's tissue, but once the skin was determined to be "alien," the rejection came more quickly. This underscored the ability of the immune system to learn; the first time the elegant defense took a while to assess the skin as foreign and build the machinery to reject it, but the second time, with the machinery in place, judgment was fast and merciless.

Eventually, Medawar turned his attention to working on transplantation in cows. Unknown to him, across the Atlantic Ocean, a key contribution to transplant science was being made in the 1950s, also in cows—specifically thanks to the story of one very lusty Hereford bull. The bull lived in Wisconsin, where, feeling its oats, it escaped its confines and mated with a cow that had already been impregnated by a Guernsey bull. The cow in turn had fraternal twins, each from a different father.

There was an oddity about these calves. Despite having different dads, they had profound similarities in their blood. In fact, each actually carried blood from the other's father. It appeared that when the calves were in utero, they had shared cell types in an unexpected way.

All this came to the attention of an immunology pioneer at the University of Wisconsin named Ray Owen, who asked why the in utero calves would accept the foreign blood of a different father rather than reject it. After all, this stepfather's blood was presumably being treated as foreign to the gestating calf. Owen floated the idea that the result of these mysterious cow couplings held some key to successful transplant and tolerance by the immune system.

Back across the Atlantic, Dr. Medawar and other scientists were making similar strides working with twin cows. They discovered that skin transplants could be made with high rates of success among twins, both fraternal and identical. What was becoming clear was that a mixture of the blood at the very early stages of life—even if the mixing involved different genetic subsets of fraternal twins and with a sibling from another father—set the stage for the way the immune system selected for alien and self. "The new science of immune tolerance was born," wrote a geneticist in a 1996 paper published by the Genetics Society of America.

Soon Dr. Medawar would go on to work in other species, eventually performing the first successful kidney transplant. (Dr. Medawar, tired of and always uncomfortable working with large animals, is reported to have said, "Thank God we've left those cows behind.")

In the Festival of Life, organ failure is, while not commonplace, hardly unusual. Livers, hearts, kidneys, and other organs succumb to disease, overuse, and injury from behaviors like drinking alcohol and smoking, to say nothing of the inevitable wear and tear of aging. So obviously it would highly limit the lifesaving possibilities if the only viable transplants came from twins. Fortunately, we've moved well beyond that.

The ultimate success of transplantation owes as much to the early experimentation of sew-and-error as it does to the discovery of drugs and the use of other tactics that suppress the immune system. The essential idea, if it's not already obvious, is to lessen the reactivity of the body's defenders so they do not attack the transplanted organ as foreign. This broadens the possible transplant matches.

Early attempts to dampen the immune system were made using radiation, but this failed (meaning that patients died) and a second wave of immunosuppression in transplants was tried using steroids. (Steroids are a class of drugs that is extremely important in the conversation about our elegant defenses and efforts to keep them in balance. I will elaborate later on the mechanics of steroids and their significance in the stories of Linda and Merredith, the women who embody autoimmunity in this book. One particular drug, cyclosporine, changed the game. The drug, approved for use in 1983, interrupts the ability of the T cell to receive an attack signal.)

The use of immunosuppressive drugs is decidedly a mixed blessing, as you might imagine. If you took this drug and contracted an infection, you would risk a muted response and serious illness. On the other hand, if you needed a new kidney to survive at all, this drug—in combination with other treatments or now more advanced treatments—would stop your T cells from full-fledged attack.

Holy cow, the human lives saved. In 2017, in the United States, there were nearly 35,000 transplants of lungs, hearts, kidneys, intestines, and other organs, according to the United Network for Organ Sharing, and by no means were the transplants done between identical twins, let alone siblings. The best possible matches are made on a number of bases, including blood type and the similarity of antigens. Even after a successful transplant, though, the recipient can need a lifelong regimen of immunosuppression.

There is another type of transplant too, one that would ultimately help save Jason's life. It is known as a bone marrow transplant, and it involves the transplantation of one immune system to another. It's the very thing that Dr. Cooper was attempting.

The possibilities for this type of transplant expanded greatly in the 1950s with a discovery that helped explain the underlying chemistry behind why one person accepts or rejects tissue from another. Work by a French immunologist, and others, isolated in human beings the first antigen that reacted against other human beings. These are called isoantigens—antigens within the same species. If two people are a poor fit for bone marrow, the isoantigens in one will provoke an antibody response in the other, setting off a defensive attack. The discovery of isoantigens earned its discoverer a Nobel Prize. This development made it possible for doctors to test ahead of time which transplant matches might be the best fits by eliminating candidates whose tissues were likely to clash. The fancy immunological term to describe isoantigens is *human leukocyte antigen* (HLA).

As a discovery, "it's major, major," a Stanford immunologist explained to me, and there's little doubt about that. It is central to the question of how the body sees "self and nonself."

Decades later, the modern version of this technology would allow Jason to receive immune cells from his older sister and help fight off a cancer his own immune system had grown helpless to confront.

But there were many more steps, as they say, between the discovery of isoantigens and Jason's bone marrow transplant. One big leap was taken by a veterinarian who helped us comprehend in a much deeper way how we (and our immune systems) understand and recognize ourselves. He found the immune system's fingerprint.

14

The Immune System's Fingerprint

For a Nobel Prize–winning immunologist, Peter Doherty is a funny guy.

He graduated from veterinary school in Australia in 1962, initially focusing on how vertebrates—like sheep (and humans)—control infection. He did so with zeal. Even in the latter half of his seventies, when I had the privilege to interview him, he talked a blue streak, an enthusiast with a sense of humor. When he was a teenager, he told me, he read Aldous Huxley, Jean-Paul Sartre, and Ernest Hemingway, and was inspired but confused. Dr. Doherty, in his own words, was a man "who was either going to go big, or crash and burn."

In his 2005 book *The Beginner's Guide to Winning the Nobel Prize*, he reflected with humor on his adolescent naiveté. "I decided to be the man of action rather than the philosopher, and resolved to graduate in veterinary science and pursue a research career," he wrote. "At this stage I was just seventeen years old, and would probably have made a very different decision if I had been more mature."

When we spoke, Dr. Doherty colorfully explained that while much had been discovered about the immune system by the time he was delving into his research, even more was still in doubt. In fact, there were still holdouts who were not convinced that there even were two main immune system cell types, the T cells and B cells. "That fact had become obvious. But some of the old guys were horrified by having to confront such complexity," Dr. Doherty told me. "They'd say that B and T were the first and last letters of *bullshit*."

As the pioneering immunologists pushed ahead, there was predictable resistance from people unsure if the course was correct. This was true with every advance. Meanwhile, the progress and pace of the breakthroughs were accelerating. It was here that science would discover the levers and knobs to permit more precision health treatment, care, and counsel. The next few chapters, before I return fully to Jason, Bob, Meredith, and Linda, will carry you deeper, joining the scientists on their journey beyond ideas and into the very molecules and systems that make you tick and that are responsible for your health. When we surface again, you'll be equipped to see more clearly the profound role of the immune system in virtually every facet of your health—both physical and emotional.

Dr. Doherty earned a PhD in 1970 at the University of Edinburgh in Scotland, where he studied sheep brain inflammation (meningoencephalitis). He returned to Australia and applied that work to mice, then started a historically significant collaboration with a visiting Swiss doctor and scientist, Rolf Zinkernagel, who had used mice to become expert at a technique for looking at the concentration of T cells when they are called into action to attack a virus.

The two scientists infected mice with a virus that can cause meningitis—an infection of the lining of the spinal cord. Then they watched as the T cells gathered around the infected cells and unleashed their fury. Most of this was actually done in a test tube.

The mouse would be infected; then its infected cells were mixed together with T cells isolated from the spinal canal.

Dr. Doherty told me that "right from the outset, our brain-derived T cells were causing the most dramatic killing anybody had ever seen.

"As disease and death guys," he added, "we were delighted!"

With further experimentation, the pair realized something crucial about the nature of this mass murder: The T cells weren't just killing free-floating infection; they were targeting mouse cells that were infected, meaning the cells that were being annihilated were part alien and part self. This was very interesting but maybe obvious. It meant the T cells were diagnosing illness inside the cell, not merely identifying freestanding viruses.

Then came a punch line—"an unexpected discovery," the Nobel Prize committee wrote in its 1996 award for the science (for work published in 1974). "The T-lymphocytes, even though they were reactive against that very virus, were not able to kill virus-infected cells from another strain of mice."

In other words, the immune system was able to discern a cell that was self and had been infected from a cell that was not self. The immune system killed only the infected ones that were self. *An individual's elegant defense didn't care simply about the infection; it cared about the infection when it attacked its own personal habitat.* Italicized because it's a key scientific insight.

If you pull back the lens, and picture the day-to-day cinema inside our bodies, what the pair of scientists had discovered was that our "killer" T cells are roaming widely to discern whether other cells that make up the tissues and organs of our bodies are normal and healthy or have been damaged in some way that's dangerous—infection, mutation to cause cancer, etc. These T cells are often considered the equivalent of hit men. But this work showed these cells have a broader function. They carry specific "receptors" that prompt them to ask questions first before they attack.

The T cells first determine if *you* specifically have come under attack. The concept is called the major histocompatibility complex, or MHC—another immunological term that goes down like cold lemonade on a freezing day.

The net consequence of MHC is that it allows T cells roaming the Festival of Life to avoid killing what Doherty calls the "normal guys" that happen to be nearby. "The assassination is precise, local, and very specifically targeted!

"The MHC is a central component of our immune surveillance system," Doherty said. "It's the key to self-recognition."

MHC is the single most varied or polymorphic of all human genes. Every human being has roughly the same MHC genes, but they are all slightly different. They are the immune system's fingerprint.

This is one of the key markers that differentiate an individual from everyone else in the world.

This extraordinary notion led to one of the most fascinating pieces of scientific theory I ran into while researching this book. The theory has to do with mate preference, incest, and the MHC.

Studies have shown the MHC gene gives off a scent. The scent is used as a factor in how people choose their mates. If one person's MHC is too similar to another's, the MHC will act as a repellent. The scent of MHC that is sufficiently different will act as a magnet.

This is significant from several perspectives. For one, it shows the unconscious drive for a certain level of diversity, given that diverse couplings provide to offspring a broader set of abilities. Relatedly, it also creates the possibility that the immune system originated not just to keep us away from pathogens but also to help us choose mates that are sufficiently self but not too much. In fact, the MHC could be part of the reason that incest has evolved to be so abhorrent.

Finally and more broadly, the role of MHC also raises a possibility that the immune system is so primitive and fundamental that it evolved in concert with a seemingly unrelated survival function: the need to reproduce. This question hasn't been answered. It is a viable theory explained to me by Dr. Thomas Boehm, a pediatrician and researcher at the Max Planck Institute of Immunobiology and Epigenetics in Freiburg, Germany.

As humans developed, he told me, "We had to make sure we did not kill ourselves by homogenizing. The one system that is ideally suited is the MHC."

In a 2006 paper, Dr. Boehm wrote: "I have proposed that this mechanism to assess genetic individuality was initially used for sexual selection and only later became incorporated into immune-defense systems. Whether this primordial system provided only transient coverage against the emerging possibility of self-reactivity and was later replaced by the MHC, or whether it directly evolved into the MHC, is unclear."

This is partially speculation. It also speaks to the likelihood that the immune system is so fundamental to human existence as to be part of the essence of the species.

I mentioned earlier that the T cell and B cell, and other core aspects of the immune system, have been in place for around 500 million years and that the foundations of our elegant defense go back so far in our evolutionary history that we share it with the world's other jawed vertebrates—a massive category that includes sharks, skates, and rays. "They have an immune system like ours, a thymus like ours, that makes T cells," explained Dr. Cooper, who has become one of the leading authorities on the evolution of the immune system.

Even as evolution led creatures to walk onto land, turned them (or us) bipedal, saw the transformation of our communications, and enabled the development of modern tools, the immune system remained largely the same. And recall that to find a different immune system (at least on this planet) you need to go back to a point in biological divergence where the jawed vertebrates split off from the non-jawed vertebrates.

This tells us that, while the immune systems are different, certain defense functions seem essential for survival. One such function is redundancy. There are multiple molecules and cells in both systems, including some proteins that seem to do virtually the same thing—whether it be attacking, inducing attack, or slowing it.

Why so much redundancy? For instance, Dr. Cooper has asked why do we need both T cells and B cells? Wouldn't one such set of specialized cells be enough? Couldn't one of the systems have evolved to take care of that? The answer to these questions remains elusive except for one basic point of proof, Dr. Cooper notes: If they weren't both necessary, they wouldn't both exist—"we don't maintain things that are not useful."

Overall, though, the scientists were getting closer, and deeper, pressing even beyond the microscopic. And with each advance came the chance to explore questions that had, previously, seemed almost pointless to ask. I'll give you one: what is a fever?

You think you know, right? I did too. The body heats up. But it's a much more profound question than I had realized, one that would illuminate a new level of understanding of the immune system, namely, that it has a vast, virtually unmatched telecommunications system. This helps explain how, when your body gets invaded, the defense signals can be sent so quickly and effectively, when necessary, calling all hands on deck.

Fever also helps explain inflammation, a concept I also figured was fairly self-evident. Not so much.

What is inflammation?

What is fever?

A stubborn scientist obsessed with rabbit fevers discovered truths previously thought unattainable.

15

Inflammation

In the late 1960s, a woman showed up at Yale University Hospital with a sky-high fever, peaking more than once at 104 degrees. In her mid-twenties, originally from the Caribbean, the woman was shaking with chills, miserable. It didn't make any sense. That's because the woman suffered no infection.

It was true that she suffered from an autoimmune disorder called lupus. But that condition wasn't known to cause this level of fever, and she otherwise had no alien infection, no pathogenic bacteria, no viruses. She harbored none of the things that are understood to cause such a fever.

This case was interesting to the doctors in general, but it was of special interest to one medical student. He became no less than obsessed. Charles Dinarello was in his third year of med school, on his way to becoming a pediatrician, and he had already developed a general interest in fever. Then he saw this young woman in the bed. The patient focused Dr. Dinarello's curiosity on fever, a subject that had long vexed researchers. It wasn't clear where fever came from or what its purpose was—to kill the infection, or was something else going on?

This is not at all a simple question. The body, for instance, doesn't have a central furnace. It doesn't have a thermostat or an organ that produces heat. But somehow, when provoked, the body—the immune system—sends its internal temperature soaring. Think for a moment about the power and strangeness of this response; the temperature rises intensely throughout the festival. Why and how?

When Dr. Dinarello started his investigation—one that would bear fruit in the mid-1970s—a few things were clear about our body's temperature, and they will sound self-evident.

Most people have a body temperature within a relatively restricted range, roughly 97 to 99 degrees Fahrenheit for adults and slightly elevated for children, 98 to 100.4, give or take. In later writings, Dr. Dinarello would note that body temperature throughout the day tended to fluctuate more "in some young women than in men." Interestingly, body temperature was at its peak at around six P.M. each day. These temperatures Dr. Dinarello referred to as low-grade fevers that are likely not representative of disease.

When we experience a fever, we become tired and develop chills. We feel it all over, a very powerful neurological response. It was one of the very earliest of medical observations—the relationship of illness and fever—with early scientists making the connection in 450 BC.

In 25 AD, one of the little-known forefathers of medicine, Aulus Cornelius Celsus, wrote of fever as one of a handful of major signs of inflammation, along with pain, redness, and swelling.

Celsus, by the way, while he was far ahead of his time, had some curious theories as to the causes of various fever-related ailments. A translation of his work produced in the mid-1930s quotes him as having written:

> Of the various sorts of weather, the north wind excites cough, irritates the throat, constipates the bowels, suppresses the urine, excites shiverings, as also pain of the lungs and chest. Nevertheless, it is bracing to a healthy body, rendering it more mobile and brisk. The south wind

dulls hearing, blunts the senses, produces headache, loosens the bowels; the body as a whole is rendered sluggish, humid, languid. The other winds, as they approximate to the north or south wind, produce affections corresponding to the one or other. Moreover, any hot weather inflates the liver and spleen, and dulls the mind; the result is that there are faintings, that there is an outburst of blood. Cold on the other hand brings about: at times tenseness of sinews which the Greeks call *spasmos*, at times the rigor which they call *tetanos*, the blackening of ulcerations, shiverings in fevers.

Pain, headache, fatigue, shivering, fever. Inflammation.
The I-word.

Yes, because of its great importance, the previous line deserves its own paragraph. A definition of inflammation written by the Institute for Quality and Efficiency in Health Care, a group funded by the German government, sums up the sheer breadth of the concept: "Inflammation is—very generally speaking—the body's immune system's response to stimulus."

In the context of health—of the lives of Jason, Linda, and Merredith, of Bob, and of you and me—inflammation is the reaction of the body to an event that challenges our well-being. This can be the inhalation of a virus, the poke of a splinter, the ingestion of noxious bacteria, the claw of a bear or a cat, or even a noise loud enough to cause injury to our hearing. At the moment of insult or stimulus, the defenses react.

Outwardly, leading signs of inflammation include pain, redness, swelling, loss of function, and heat, including fever. Each of these flows from activity going on inside the body aimed at limiting damage from the insult, and also repairing the damaged area.

Before I turn to fever and its discovery, I want to put fever into context by illustrating the broader inflammatory picture.

Say you step on a splinter. Virtually instantly, your body recognizes the need for a response. As a preparatory step, the blood vessels in the area open, or dilate. This allows more defenders to reach the action, and it leads to redness and heat in the region. More blood, more cells, more oxygen. The blood vessels go through a second change, becoming more permeable. Now other defenders can move into the tissue, along with clotting agents. These are different kinds of proteins, and as their numbers grow, the region experiences swelling. All this activity can lead to pain. In that way, the inflammation can be seen as having an important impact on behavior by, say, limiting use of the foot that stepped on the splinter so that your elegant defenses have time to shore up the skin.

The inflammatory response, aimed at making sure the area of invasion is fully secure, can well exceed the pinprick of impact. In fact, there can be more tissue damage twenty-four hours after the insult takes place than there was at the moment it happened. During that period, the elegant defenses are examining, cleaning, and rebuilding enough of a physical space to make sure that no danger is left behind and that they can rebuild healthy new tissue seamlessly in the region around it.

Another example of an everyday inflammatory response is the one set off by the common cold. It often is caused by a rhinovirus, which makes a battleground of your nose. The virus replicates there. The blood cells in the area open up to allow for easier access by immune cells. They flood in. Swelling! The vessels become permeable, allowing more flow of fluid. Leaking! Your stuffy nose explained.

So what does this inflammatory response look like up close, on the molecular level?

It resembles the aftermath of a disaster—an armed attack, a multicar crash, a hurricane. I distinguish such events from, say, a fender bender, where one cop might show up and send everyone home. When an insult, like the splinter, happens, it might seem like a fender bender from the outside, but our elegant defenses need a lot of information to make that call and to repair the area of the insult, however small it may be, and this brings multiple cells into the fray. Let's meet them.

I've already introduced you to one of the key cells. It is called a macrophage. It is the cell type observed a hundred years ago by Élie Metchnikoff, the Russian scientist who, as I described earlier, stabbed a starfish larva with a splinter and, looking through the microscope, saw wandering cells swarm to the site of the insult. Metchnikoff observed macrophages devouring other cells in the region of the splinter.

The technical term for one cell eating another is phagocytosis. That word derives from the Greek word *phageîn,* which means to eat. So the macrophages are big (macro) eaters. These cells are like the love child of a janitor and a cop who eats first and asks questions later. They attack cells in the region that might be damaged or infected by consuming them and then chemically blitzing the devoured particles.

These macrophages derive from and are a subset of a broad group of immune cells known as monocytes. Some monocytes turn into macrophages. But others take on a very different function.

Until now, I've described largely T cells and B cells. If you're surprised to hear there are any more cells in the immune system, you're in good company. In fact, the more that immunologists over the previous century explored inflammation, the more they realized that our defenses are broken down into many different cells and receptors with widely varied functions. This reckoning ultimately forced them to redefine the very nature of our immune system, though not until the late 1980s.

In the meantime, piecemeal discoveries were made that were crucial in determining the different types of cells and their functions essential to the body's defense.

For instance, I mentioned that macrophages are a kind of cell called a monocyte. Then in the mid-1970s, Ralph Steinman threw a wrench into science by discovering a sibling monocyte.

"In science, it is a rare event for one individual to make a discovery that opens a new scientific field, work at the forefront of its research for forty years, and live to see his endeavors transformed into novel medical interventions. [Steinman's] discovery of dendritic cells changed immunology."

Thus began the Nobel citation about the work of Dr. Ralph Steinman, a physician and researcher who in 1973, trying to fill out the details of an immune system that looked increasingly complex, used an electron microscope to discover an unusual-looking cell. The cell had long tentacles resembling branches, hence the name derived from *dendron,* which is the Greek word for tree.

Dr. Steinman and a collaborator surmised that these cells played a key role in the immune system, and then proved it. Through a series of experiments, they showed that these cells, when presented with a foreign cell or organism, could stimulate or induce a powerful response from T cells and B cells.

Steinman's research began to get at how these dendritic cells worked. He showed that the treelike cells played a key role in presenting antigens to the immune system cells. This allowed the

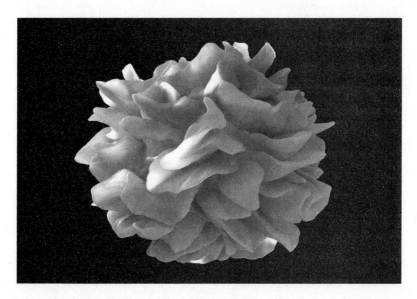

Artist's modeling of a dendritic cell. (NCI/NIH)

T cells, for example, to see whether their receptors fit the antigens being presented.

In practical terms, when your body is invaded by an alien organism, the dendritic cells take a piece of the organism and display it to soldiers and generals to determine if attack is warranted. The dendritic cells roam the Festival of Life, brushing up against guests at the packed affair, and presenting their identities to the T cells. If an antigen were perceived as foreign, it would stimulate a heavy response, what is known as a mixed leukocyte reaction, or MLR, a major inflammation of T cells and B cells and other immune cells.

Some scientists initially dismissed this discovery. It seemed to differ from or even contradict the commonly held belief that macrophages were a kind of front-line immune cell, largely distinctive from the all-powerful T cell and B cell. Bit by bit, though, evidence mounted that the T cell and B cell are being tremendously aided by, and are reliant upon, other cells. In fact, T cells and B cells, together known as lymphocytes, make up only as much as 40 percent of the white blood cells.

The monocytes comprise 5 percent, give or take.

The biggest chunk is made up of cells known as neutrophils. They are both spies and assassins.

The neutrophil is the cell that Metchnikoff had observed and had himself eventually better understood. The neutrophils represent more than half of our white blood cells—50 to 60 percent. Their work in the body, we now know, is a bit like that of a cold war spy—a deadly agent but one who is often quietly looking and listening for trouble and then occasionally getting drawn into violence. The journey of the neutrophils begins in the bone marrow, where the defenders are born and from which they make their way into the bloodstream and circulate. The neutrophils might dip into tissue or an organ for a bit, look for pathogens and, finding none, then return to the bloodstream, to continue monitoring and *smelling*. They can pick up scents, or chemical releases, of pathogens.

When they "smell" such a thing, the neutrophils squeeze from a blood vessel into the tissue where the infection has taken root. The neutrophils, drawn magnetically to the infection, begin to eat it, devouring the invaders. Then the neutrophils release a chemical called an enzyme, which destroys the pathogen. It is a violent affair, one that leaves the neutrophil itself spent, almost like a bee that has inflicted its one sting. The neutrophil begins to dissolve, creating of itself digestible chunks that can be cleaned up by cells with a more janitorial function.

In the Festival of Life, neutrophils are first responders.

"If you scrape your hand right now, and get an infection, the first cell there would the neutrophils. The macrophage comes soon thereafter," said Dr. Anthony Fauci, whom I mentioned earlier, the director of the National Institute of Allergy and Infectious Diseases at the National Institutes of Health, and one of the most influential contemporary scientists. His story will eventually become closely intertwined with that of Bob Hoff, the man who fought HIV to a draw.

In much smaller concentration in the body are two other defenders, the eosinophil (less than 5 percent of the white blood cell population) and the basophil (less than 2 percent). Combined, they are known as granulocytes. This name reflects their function. These cells contain tiny enzymatic granules that digest and destroy pathogens.

In the 1970s an experiment involved yet another immune cell—the natural killer. The discovery is interesting in and of itself, and also because it was part of the broader reframing of the understanding of the immune infrastructure. The scientific narrative had been about the primacy of the T cell and the B cell. That narrative was falling apart.

The story emerged in 1975 in a paper titled "'Natural' Killer Cells in the Mouse," published in the *European Journal of Immunology*. It described an experiment that didn't seem to add up.

The study involved mice raised in antiseptic environments such that they had not faced a challenge to the immune system. As a result, it was not possible that the immune systems of these mice would've had a chance to learn and respond to a particular threat.

Cells from the spleens of these pure mice were extracted. They were introduced in a test tube to cancer—specifically, leukemia cells.

The strangest thing happened. There was an immune reaction. The spleen's immune cells attacked. That alone wasn't necessarily at odds with previous learning; after all, perhaps the immune cells had antibodies that recognized something foreign. But, quite oddly, the attack didn't involve any B cells or T cells. The response was less specific than the targeted nature of B cell and T cell attacks. These "new" cells swarmed instantly in a kind of raw, generic manner that seemed more consistent with a knee-jerk attack than the specified, deliberate nature postulated by clonal selection theory.

This was different, and possibly hugely important. But what were these things?

The scientists called them "natural killer cells." They seemed to belong to the same family as T cells and B cells, but they behaved very differently.

"The natural killer cells did not get much respect when they were discovered," reflected David Raulet, an expert in the field from the University of California at Berkeley. "A lot of people working on T cells sneered at them and didn't think of them as relevant."

The authors of the paper themselves acknowledged an oddity. In a summary, they wrote that "the spontaneous" attack of the mouse spleen cells "is exerted by small lymphocytes of yet unde-fined nature."

One reason scientists had trouble absorbing the new information was that, just as the human body itself struggles with the alien, science can have trouble making peace with *ideas* that seem for-eign. Scientists and thinkers with entrenched theories rejected the challenge to the prominence of T cells and B cells, as if these new discoveries were alien tissue or pathogenic bacteria. Ideas, memes, can elicit a kind of autoimmune response, an overreaction that feels protective initially but can ultimately prove counterproduc-

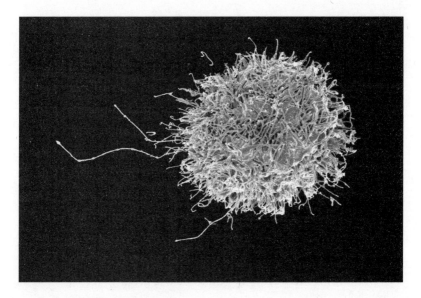

A natural killer cell. (NIAID/NIH)

tive and make it harder to find truth. (On the other hand, kudos to immunology for finally giving a name—natural killer cells—that was accessible, described what the cells actually did, and that Madison Avenue would at long last have lovingly embraced.)

How did this litany of new cells fit together? What was the interplay?

"We couldn't connect the dots," Dr. Fauci said. How did all these things work together?

A partial answer came from the mid-1970s revelation about fever—a discovery set off by the temperature spikes exhibited by the Caribbean woman who showed up at Yale and that became the obsession of Dr. Charles Dinarello.

16

Fever

"For centuries before the introduction of the thermometer, fever was a well-recognized sign of disease," Dr. Dinarello wrote in 1978, as he was changing the world of immunology. "Only during the past three decades has the mechanism by which disease causes a rise in body temperature begun to be clarified."

Dr. Dinarello dates the relevant research to 1943, when a Russian scientist who had relocated to the United States found that he could induce fever in rabbits by injecting them with pus. Pus, it turns out, is the detritus of neutrophils, the cells that rush into action at the first sign of insult. They kill what is around them and die in the process. When you observe pus oozing from your body, you're seeing these dead cells.

The 1943 paper posited that the fire was sparked by neutrophils. This was wrong, but it was a start.

Why rabbits? Rabbits make good guinea pigs for science because they can be somewhat trained, and changes to their behavior are relatively easy to observe.

Early on, it was discovered that injecting rabbits with pus could elicit a fever. It was a first step in looking for the exact pyrogenic, or fever-producing, process. Through the 1950s and '60s, more evidence was added about the process. For instance, the rabbits conserved heat during fever by constricting blood vessels, such that their ears became cold. (Have you ever felt clammy when you

have a fever?) "The rabbit becomes quiet and motionless," Dr. Dinarello wrote in a history. "This observation resulted in the discovery," he added, that the pyrogen "was a sleep factor."

Then in 1967, science got closer with a surprising finding. A paper published in *The New England Journal of Medicine* reported evidence of a pyrogen—a fire starter—in a blood cell different from
the neutrophil. Rather than coming from a first-responder killer,
the chemical that seemed associated with fever derived from a
monocyte, which is a kind of macrophage. Understandably, it had
been hard for earlier scientists to tease out these cells. So which
was it, neutrophil or monocyte, and what difference did it make
anyway?

This was largely the state of affairs when Dr. Dinarello observed
a woman at the Yale hospital running a high temperature who
shouldn't have had a fever because she did not have an infection.
The case was intriguing, and he was already interested in fever. "I
said, 'Goddammit, I'm going to discover what this molecule is,'"
he reflected. He aimed to solve the fever riddle.

Dr. Charles Dinarello—don't call him Charlie–grew up in a Boston suburb that, as he put it, was filled with Italians, Jews, and
Irish. His grandparents were immigrants from mainland Italy and
Sicily. His mother never finished high school, and his father was
blue collar. Charles wound up, as you now know, at Yale Medical
School and finished with a prize for having the most outstanding
thesis. It was about fever.

The Vietnam War was raging, and as a medical school student,
he'd had the same choice as his med school peers: sign up for government research or risk getting sent to patch up land-mined boys
in the helicopter zone. The choice wasn't quite that simple, but it

did feel to many young doctors that government work in Washington would protect them from a combat zone. Dr. Dinarello went for research and wound up working at the National Institutes of Health. Not only that, he earned his way into a remarkable place and time: Building 10 at the NIH, a hall of truly great science, a Willy Wonka factory of experimentation and discovery.

It's a huge flat-faced brick building, set on a campus that is part of the largest clinical research center in the world. Patients and scientists. Collaboration. It represents an extraordinary commitment to science by the United States and by President Dwight D. Eisenhower. Between 1950 and 1960, the NIH budget grew from $53 million to $400 million. The funding was largely a bipartisan affair, although resistance came from some Republicans wary of expanding government. Nothing like the battles seen today, though. And as history will bear out, the science done at the NIH would go on to save many lives, including, very arguably, Jason's. Seeds of salvation for sufferers of cancer, AIDS, autoimmune diseases, flu, and other killers were planted in Building 10. The work done here speaks to the power of a broad field called basic science, defined as science that is aimed at understanding core concepts and isn't directed at developing, say, a particular medicine to attack a specific disease. Basic science is more diffuse, an act of faith and failure—many projects don't work out—but the sum of the effort has been the lifeblood of cures for many major diseases.

Dr. Dinarello's lab was located on Building 10's impressive eleventh floor, at a time when immunology work there was exploding. The eleventh floor wasn't impressive for its confines—it was a mess—but rather for the brain power there. Around every corner sat an ambitious, bright, creative thinker.

Dr. Dinarello was easy to pick out. He was the one with rabbit shit under his fingernails. It was from digging around down there with the rabbit rectal thermometer.

"I'm joking," he told me. Sort of. "But the reality is that I did have rabbit feces under my fingernails for twenty years."

It was now 1971, and his first task was a bureaucratic one. He had to convince fellow researchers and his boss (a luminary named Sheldon Wolff) that he should be allowed to hunt for the body's own fever molecule. Some were dubious. Could Dr. Dinarello be sure, for instance, that he'd filtered out every other molecule, and not only that, could he be *absolutely sure* that the cause of the fever wasn't an alien substance, an infection?

Consider for a moment the profundity of that question. There had long been an assumption that fever was linked to infection. In contrast, what Dr. Dinarello was pursuing was the idea that infection need not be present, and that, in a manner similar to the case of the woman with lupus he'd witnessed in medical school, the body was generating the fever with its own molecule, without necessarily being prompted from the outside.

Eventually, he got his wish to pursue the project, and then he ran into a very practical problem. Where was he going to obtain the white blood cells? "Where to get billions and billions of monocytes every day? That's when the project would get serious. It was an important step," he told me. He has the knack for spinning a yarn, and I could hear him picking up steam. A quick reminder: Monocytes are roughly synonymous with macrophages. The difference is that monocytes are immature macrophages. When these cells come out of the bone marrow, they are monocytes for a few days until they diffuse into the tissue and then become macrophages. For the sake of simplicity, and without losing any accuracy, I'll just say that Dr. Dinarello suspected macrophages were involved, but he needed a bunch of them.

That's when he discovered the trailer.

It was out in the parking lot of the NIH. It had been put there to experiment with a new technology that involved giving blood platelet transfusions to cancer patients being treated with chemotherapy. The provision of all these platelets involved using a lot of blood. The white cells weren't of interest to the folks in the trailer.

"I'd go there every late afternoon and salvage these cells. Just take them, in a blood bag."

The rabbits were furry white. "I treated them like they were my kids," Dr. Dinarello said. He would train each rabbit for two weeks so that it would be calm when it underwent the procedure. "After a couple of weeks, they were ready," he said.

To prepare the environment and the macrophages for injection, he was meticulous about the surroundings. "I stayed away like the plague from any bacterial product that would cause fever. I couldn't risk any contamination." He knew that his experiment would be rejected if his peers suspected that the cause of fever was an antigen or a bacterium.

Dr. Dinarello took the white cell "dreck" from the trailer. He then mixed these immune cells with dead staph infection to stimulate a macrophage reaction. He injected the mixture into rabbits, knowing the experiment would provoke a response in his furry friends.

As he tells the story, he pauses, as if struck by the oddity of his obsession. "It took six years to purify this molecule. If you ask me what drove me—why not give up and take on an easy project?—

I'll tell you: It comes from observing the physiological change in this rabbit—to see a rabbit get still, its ears get ice cold. Within ten minutes, there's this horrible, dramatic thing. I had to know: What is this molecule doing to the brain?"

Four years into his six-year quest, he was interrupted. He had to fulfill a commitment to be chief pediatric resident at Massachusetts General Hospital.

He returned in 1975. By then things were exploding in immunology around the world, with new technology allowing new techniques. One involved radioactive labeling to help identify, purify, or cull out individual molecules. In Building 10, down just two floors on nine, was a guy who was pretty good at the technique. His name was Christian Anfinsen. He'd already won the Nobel Prize, in 1972. Dr. Dinarello asked if Anfinsen might not help close the deal on the rabbits and their fire starter.

Closer and closer they got, homing in on a purified molecule, isolating it from other contaminants and molecules. And then, one day in 1977, something strange happened. The molecule disappeared.

This was the moment, the revelation. When the molecule disappeared, Dr. Dinarello realized the fever-inducing molecule was so purified it only looked absent. Of equal importance, he'd discovered that the amount of that molecule could be virtually nonexistent and still light the body on fire. The importance of this is hard to sufficiently emphasize. It takes very little of this thing to cause a major reaction in the body.

"It is perhaps the most important statement of my career," he said. In technical terms, he's referring to the discovery that it requires an amount as little as ten nanograms per kilogram of this substance to start fever. In translation: "It was a thousandfold less than anybody ever predicted. It was amazing. This molecule is so potent."

And it came from a monocyte, one of those immune cells like the macrophage (which devours refuse and pathogen), but one that now appeared to have much broader function. Dr. Dinarello called it a leukocytic pyrogen—a fire starter born of the white blood cells, the leukocytes.

"He realized, 'Oh my god, it's not coming from the neutrophil. It's coming from the monocyte,'" recalled Dr. Fauci, who worked with Dr. Dinarello on the eleventh floor. When Dr. Fauci related the story to me, his own voice rose with elation. His excitement took a moment for me, an outsider, to appreciate, given how thick these conversations are with immunology speak. But the emotion broke through. This was huge.

Dr. Dinarello published his first paper in 1977. His revelation was initially hammered. "The Germans wrote papers against it," he said. "They said. 'He's got contamination.'"

Slowly, though, the reality sank in.

In fact, in Ermatingen in 1979, Switzerland hosted the Second Lymphokine Workshop. The assembled, having accepted the notion, decided to give a new name to these so-called mediators. Henceforth, a leukocytic pyrogen would be known as an interleukin. *Inter* from a root for "means of communication." *Leuk* from the Greek root for white, as in *leukocyte* (white blood cell).

Broadly, the leukocytic pyrogen was a kind of mediator, communicator.

Interleukin-1, the first interleukin, was born. Dr. Dinarello might fairly be called its midwife. Armed with this knowledge alone, you're on your way to earning a bachelor's degree in immunology.

The story doesn't quite end there. Maybe the most important part was yet to come, and it would make Dr. Dinarello quite a controversial figure.

One Saturday morning in the mid-1970s, there on the eleventh floor of Building 10, Dr. Dinarello was working with another scientist, playing around with his purified molecule. They wanted to see if interleukin-1 had any impact on the larger immune system. Did it do something besides stimulate fever?

In rough terms, the experiment involved giving a deadened human virus to a rabbit, stimulating interleukin, injecting that product into a mouse, and looking for a T cell reaction. To measure the reaction, they then went into the "counting room," where the machine that measured the radioactive labeling would click, like a Geiger counter, when measuring a particular molecule or cell.

"We're watching every two counts to see if the T cells are activated. All of a sudden, the counter went berserk. *Ba-bid-a-ba-bid-a.* It was like a science fiction movie," Dr. Dinarello related. Another scientist in the room was the one who had been working with the mice and T cell stimulation. As they watched the clicker go wild, indicating a massive increase in T cells, "Lanny says to me: 'What the hell did you give me?'" Dr. Dinarello said. "'This is a million-fold more active than I've ever seen.'"

What did this mean?

At the most basic level, it meant that the interleukin-1 was inducing not just fever but also a T cell response.

So what?

Recall that much of immunology was still focused on the prominence of the T cell and B cell, but particularly the T cell as the commander in chief in the alliance. Now, though, it looked like the macrophage was prompting the T cell, not the other way around.

"From 1976 to 1979, I was scared shitless to publish it," Dr. Dinarello said. "How can a molecule produced by a human monocyte that causes fever in rabbits also cause a lymphocyte reaction in mice? It was heresy for immunologists." Dr. Dinarello's idea, which ultimately would be proven correct, goes to the heart of how we now understand the immune system, and also how we've come to try to manage and even manipulate it—in cases like those involving Jason, Merredith, Linda, and Bob.

This era, which is still ongoing, involves the discovery of dozens of powerful molecules that show the extraordinary complexity of our elegant defense, the multiple actors with overlapping duties, wonders of science as strange as any fiction. Enter Flash Gordon.

17

Flash Gordon

In a 1960s Flash Gordon comic, doctors on a spaceship used a wonder drug called interferon to cure a patient on the verge of death. Flash Gordon was fictional. The drug was not. The concept had emerged several years earlier when two scientists, one Swiss and one British, made a curious observation while experimenting with viruses and baby chickens.

The scientists took a virus from chicken eggs and killed the virus in an acid bath. Then they added this "deadened" virus to another chicken egg and then added live virus. The live virus didn't grow. The deadened virus had interfered with the development of the live virus.

Hence the name: interferon (IFN).

The scientists' theory was that the healthy cells had picked up a signal from the inactivated virus that deterred growth. Did this mean that some message had been sent blaring: *This is an inhospitable environment, so don't waste resources here*? It wasn't clear how interferon worked, or even exactly what it was.

Immunology became increasingly enamored with the idea that it might isolate and corral the messaging system. What made this notion so significant is that it entailed using a natural substance to fight disease. The alternative, building medicines around foreign substances, almost invariably provokes side effects because it stimulates interest from the immune system and causes inflammation.

Or consider the horror of chemotherapy, in which terrible toxins attack tumors but at the cost of scorching self.

Imagine if, instead, a harmless dead virus—a wholly natural and innocuous compound—could be harnessed to stop deadly live viruses. The promise inherent in that possibility grew as microbiological technology improved and scientists could see that one of the key properties of interferon was that it prompted the activation of genes that produce chemicals that attack viruses. Also, in the 1970s, it became clear that interferon, identified by now as a protein, had several subtypes. Maybe then it could have broad applicability.

Indeed!

There was a period (speeding briefly ahead) during which drugs built around interferon had a market value worth tens of billions of dollars, though it is typically no longer a front-line treatment. Diseases like hepatitis were treated with drugs that entail an injection of interferon in combination with ribavirin. The interferon bolsters the body's own defenses by sending a message to the immune system to attack the virus.

But to get to this point (speeding back in time now) required scientists to purify interferon, a key step not unlike the challenge of purifying interleukin. Enter what until this point had been a kind of foreign organism in the world of immunology: a woman.

Her name is Kathryn Zoon, part of a generation of women shattering gender barriers in science and expanding the definition of "self" in a field long dominated by men. In 1966, at the Rensselaer Polytechnic Institute, Zoon held the distinction of being the only female chemistry major in her class. She was one of only a few women at the prestigious technical school—"a rare bird," she said. Fellow students seemed largely unfazed, including her future husband. Not always the flummoxed male teachers. "Some of them just wouldn't even look you in the eye," she recalled.

Her merit prevailed. At her graduation, she won the prize for being the top chemistry student.

By the mid-1970s, the world of science was at last changing. Zoon went on to obtain her PhD in biochemistry at Johns Hopkins in 1976 and then was accepted onto the ninth floor of Building 10 at the National Institutes of Health, where Zoon joined the lab of Christian Anfinsen. It was Anfinsen who had counseled Dr. Dinarello on chemistry techniques to isolate interleukin.

Dr. Dinarello on the eleventh floor worked with rabbits. For Zoon on nine, the guinea pigs were sheep. Sheep, you don't keep in the lab. They were housed on a farm in Poolesville, Maryland, about a forty-five-minute drive from Building 10 in Bethesda. It was a veritable zoo, with mice, sheep, monkeys, and, yes, rabbits too. A courier would drive from Building 10 every few days to the farm with a deadened human virus.

At the farm, vets injected the partially purified interferon into the sheep. Then the vets would withdraw plasma, including white blood cells, from the sheep. The idea was that the sheep plasma included antibodies that had reacted to the interferon, and these antibodies were used in the purification of the interferon. Then once the IFN was pure, Zoon and her colleagues, including collaborators at Caltech, sequenced the interferon.

It took four years, but in 1980 they released a paper describing the pure form of interferon, allowing the substance to be manipulated, tested, and turned to medicine. Eventually researchers would identify three types of interferon: alpha (A), beta (B), and gamma (G), and then much later, lambda (L).

The role of these took a long time to fully understand. But it's worth skipping ahead to fill in the significance and role of the mighty, tiny secretion that is interferon A, a family of twelve related proteins.

"It's the first step that our bodies have to deal with a foreign agent, a virus, or a tumor. It's the first line of defense," Zoon told me.

I can sense readers' puzzled looks, an eyebrow raised over your book or gadget. Haven't you already read in this book that some other cell or substance is the first line of defense?

Yes, you are correct to raise an eyebrow. You are not missing something. The thing is, the immune system has multiple overlapping, sometimes redundant first lines of defense, and second lines too. This festival—our take-all-comers cocktail party—is nothing if not chaotic and multifaceted. There also is method to the madness. Multiple actors roam the party, using different tactics, often overlapping.

Not just that. "A lot of different cell types can make interferon," Zoon explained.

Say, for instance, a virus seeps into your nose or slips down your throat. The invader interacts with a healthy cell. That cell detects molecules consistent with a foreign agent. Within that tiny cell, a kind of supercomputer-like process starts that leads to changes in proteins and, in turn, the secretion of alpha, beta, and lambda interferons. Or maybe the cell dies from the invasion, but before it succumbs, it manages to go through the protein changes to create interferon. Other surrounding cells pick up the presence of interferon.

"This starts a chain reaction," Zoon explained.

It can cover an isolated region—say, an organ—or spread through your entire body within just a few hours. Cell after cell begins to pick up the signal and create interferon and other proteins that protect the cell. Once it does so, the interferon, true to its namesake, induces the manufacture of proteins that interfere with the ability of the virus to reproduce itself.

It comes with a side effect.

"When interferon is secreted, you feel sick. It causes aches and pains; you feel terrible," Zoon explained. Your behavior is being modified—*not by the virus directly, but by the response.* In very practical terms, a virus invades. The early warning system then sets off a cascade that leads to inflammation, and that also makes you feel rotten. Tired, sore, hot, as I described earlier. You are slowed down, and this can have the very beneficial impact of diverting your body's resources to fighting the virus and not, say, focusing on your job or going for a jog. Your defense system needs your limited energy.

Your immune system takes care of you partly by making you take care of yourself. And it would be tempting to say without reservation that your feeling rotten is a sign to withdraw and let your body heal. But it turns out the practical side of this is much more complicated. This is where the forthcoming stories of autoimmune sufferers Linda and Merredith become instructive. Sometimes the immune system overreacts, while other times, it is beneficial to push through feelings of sickness to reduce the inflammation. For more on this, stay tuned.

Their stories become more accessible, more meaningful, with additional core science. Interferon belongs to a broader set of chemicals that prompts immune system action. This set of chemicals informs virtually all of disease, including how we respond to it.

Meet the cytokines.

A cytokine is a secretion from a cell that prompts action by other immune cells. It is a messenger. It can be sent by an interferon or any of a number of other immune system actors. In the Festival of Life, when a foreign agent bursts into the party, immune cells might send lots of cytokines to one another—pulses of communication.

This puts a fine point on a major concept in the understanding of the immune system: It has a telecommunications network. Full stop. Our defense network is sending signals across the body. In the case of fever, signals wind up in the brain, at the hypothalamus, a neurological region central to temperature regulation. Then the signal travels through the body, calling on other cells to stimulate a fever. Interferon works in a similar fashion.

The immune system's communications network rivals in power, speed, and reach any communications network the world has ever invented. (Take note, Silicon Valley!) The monocytes call out across the body's galaxy. And do so without wires and across vast distances millions of times greater than the actual cell itself.

"These telecommunications are essentially wireless. One cell doesn't have to touch another," said Dr. Fauci. The system "is plastic, flexible, and enormously complicated.

"It's like a supercomputer."

It's worth pausing to think about how far immunology had come since the late 1950s when Dr. Miller discovered that the thymus wasn't just a waste of space, or God's throwaway line. The thymus makes T cells. The bone marrow is the origin of B Cells. They flow in the tunnels and vessels that make up the lymphatic system and congregate in lymph nodes and lymphatic tissue. These are like command centers, surveillance hubs where the firefighters are awaiting a call. The T cells, when alerted by dendritic cells, behave as soldiers and generals, spitting out cytokines; the B cells use antibodies to connect to antigens as if they are keys in search of a lock. Macrophages, neutrophils, and natural killer cells roam the body, tasting and exploring, killing. These networks get connected by signals, chemical transmissions, or processes; are spurred on by interferon and interleukin; and can induce powerful side effects, like fever.

Conceptually, this is the kind of cascade that keeps you healthy. The system goes after parasites and viruses, bacteria and malignancies. It works nonstop, picking up minor threats that we never experience on a conscious level, and midlevel threats that send us to bed, and myriad major threats that might well kill us absent the presence of this system. In an historical sense, I've described a complex system—at least compared to what science understood in Dr. Miller's day.

The stage was set, through science and the technology that supported it, to discover lots of different molecules and cytokines. Once there were only T cells and B cells, and then suddenly, there was a laundry list of molecules monitoring and policing the Festival of Life. Their individual discoveries came with a revelation about their overall purpose. Some, of course, are involved in identifying and attacking outsiders, but many others monitor our own immune system to make sure that it doesn't overreact. Together they are the interleukins, known as IL for short. They roam the Festival of Life, checking for outsiders, inspecting each other.

For example:

IL-1 induces fever.

IL-2 causes T cells to grow.

There's IL-6. That causes B cells to grow.

IL-2 and IL-6 are powerful ones, with a twist. The problem with these interleukins is that they can become too abundant, their signals too aggressive. That leads the body to attack with too much ferocity. This is called autoimmunity. Even if you've never experienced the dramatic, chronic challenges faced by people like Merredith and Linda, you've surely in your own life felt the impact of your immune system firing too aggressively, causing you, for instance, to feel fatigued when you'd be better off getting off the couch and walking, or to experience pain that has no apparent external cause or a hint of fever.

If left unchecked, the threat from autoimmunity is no less than deadly. That's why our immune system has evolved to have its own system of checks and balances. In fact, many interleukins are designed to be anti-inflammatory. They are immune system brakes, not accelerants.

In fact, some of the sets of monocyte cells that help fuel inflammation also have subsets that dampen inflammation. For example, we now know that the IL-1 family has dozens of members, of which many are anti-inflammatory. At least a third of the variations of this key principal immune system protein are designed to stop the immune system from inflaming.

"Before antibiotics, these inflammatory cytokines helped kill off infection," Dr. Dinarello says, and the cytokines still play that role. How do the cytokines know to turn off? What happens if they don't turn off? "If you fail to make anti-inflammatory cytokines, you die of mild inflammation."

That's how powerful this system is. Mild inflammation, wholly unchecked, can kill. Dinarello likes the analogy that the immune system has turned the body into a police state. "You need inflammation to protect against invaders. You need policemen. But if police get too rambunctious, they cause damage and kill innocent people."

The discovery of all these proteins provides evidence of what Dr. Fauci told me so eloquently about the immune system. It's a supercomputer.

Dr. Fauci was poised to redefine its purpose.

18

The Harmonious Way

The year was 1980, and Dr. Fauci was a rising star, eventually one of the brightest lights in immunology. Since 1972, he'd been on a quest to figure out how to deal with what he calls "aberrant" immune system responses. He meant situations in which the immune system attacks the body.

He'd done extensive pioneering work on medicines that help dampen the immune system when it attacks the body. "We had to calm down the immune system by suppressive agents without necessarily suppressing it so much they were susceptible to infection," he said.

During this period, Dr. Fauci hadn't put so fine a point on it, but he was helping define a new identity for immunology. For many years, the field had viewed the immune system as something poised to "attack, seek, and destroy."

Dr. Fauci could see that this was just half of the equation—in fact, well less than a full definition.

At its core, what the immune system was doing wasn't simply seeking and destroying. Instead it was looking for a balance—between attacking and neutralizing real dangers and showing sufficient restraint such that its potency didn't destroy the body. In 1980, Dr. Fauci helped capture this pivot in immunology by naming a new lab at the NIH. He called it the Laboratory of Immunoregulation.

Mark the moment. The story of the immune system became the story of homeostasis—a state of harmony or stability. This is what makes our defense so elegant. It is a system precisely and delicately tailored to stay in balance, keep the peace, and do as little damage as possible to us and our surroundings.

This balance is central to our health, as you'll see momentarily in the lives of four individuals you'll soon meet again—Bob, Linda, Merredith, and Jason.

First, though, I will introduce you to three wise men and a discovery that turned the science of immunology into healing medicine. This was the point of practicality for the long-opaque world of immunology, a turning point where the decades of science became lifesaving treatment.

19

Three Wise Men and the
Monoclonal Antibody

"It's a story that revolutionized science and medicine," writes Dr. Sefik Alkan, a Turkish-born immunologist and historian. The discovery is now used in diagnosis and treatment of the pantheon of diseases "from rheumatoid arthritis to cancer."

Now we're getting close. The pieces are coming together, the exploration leading to application, to real-world solutions. None, arguably, was as significant as the discovery of the monoclonal antibody. This next scientific treasure likely will touch every reader at some point, if not directly, then through a family member. So it's useful to grasp this piece to understand what might someday be injected into your body to extend or save your life.

The story starts like this: A Dane, an Argentinian Jew, and a German walk into a research lab . . .

The first of the three wise men was Niels Jerne, a Danish immunologist who was among the elite thinkers of his era and the founder of the Basel Institute for Immunology. "In his office," Alkan writes, "there was a long table adorned by dozens of scientific journals; all were being read regardless of language (English, Dutch, Danish, French, and German)."

Jerne had created a way to isolate and count antibodies.

The discovery here is referred to as the Jerne plaque assay. From the University of Windsor website, I'll draw the first few steps in what I think of as a kind of recipe—a dip of the toe into the complexity of immunology—and then I'll just summarize the darn thing and its meaning.

1. Put 2.0 ml of Hank's balanced salt solution (HBSS) in a small mortar and cool it in an ice bath.
2. Kill the mouse with an overdose of ether by placing the mouse in a small jar with an ether-soaked cotton swab and replacing the lid.
3. Remove the dead mouse from the jar, put it on a paper towel, swab the abdomen with 70% ethanol, and cut open the abdomen. Cut out the spleen and make sure that excess fat and tissue are removed. . . .
4. Put the spleen in the 2.0 ml of cold HBSS and cut it into small pieces. Grind these small pieces with the pestle until an even cell suspension is formed.
5. Filter the suspension through a cheesecloth that has been placed in a small funnel. This will remove any large clumps of cells. Flush the few remaining trapped cells from the cloth with 5.0 ml of cold HBSS.

You get the idea of its complexity (which eventually involved a centrifuge; more salt baths; mouse spleen cells, having been washed, put onto slides, sealed with paraffin wax, then incubated; and finally, the viewing of the results under a microscope).

What resulted was a plaque that, viewed under the microscope, would allow the counting of antibodies.

It was a huge step. Why? When you contract a virus, your body generates antibodies to fight it. Thanks partly to Jerne, doctors regularly use tests that isolate our antibodies as a way to understand the type of bug we're fighting, how effectively we're fighting it, and the intensity of the fight going on between our immune system and the pathogen.

Wise man number two was César Milstein, from Argentina. He had figured out an ingenious way to create lots of antibodies for purposes of studying them. His tactic for generating antibodies involved mating a B cell with a cancer cell. This worked wonders because cancer cells, for all their evils, have an important scientific value: Cancer cells grow and grow. They are the body's weeds. What Milstein did by fusing a B cell with blood cancer, called myeloma, was to create a lineage of B cells with cancer's powerful reproductive cycle. Now Milstein had a petri dish filled with antibodies, which allowed science to study and experiment with huge batches of these precious defenders.

In 1973, Milstein came to Basel to give a talk on this process, and listening there was scientist number three, Georges Köhler, the German.

Long (and complex) story short (and simple), Köhler combined the techniques of Jerne and Milstein. He used mice and sheep to isolate individual antibodies and then make countless copies of them.

For the first time, scientists could isolate a cell with a particular antibody and make endless copies of it. In turn, this technology allowed researchers to begin to make distinctions between and among lots of different cell types with antibodies. This was akin to creating the most powerful microscope that cell biologists had ever seen because it let them distinguish one cell type from another, determine which had what kinds of antibodies and also how many antibodies appeared on each different cell.

As a first basic step, this began to reveal that, for instance, B cells were far more varied than people originally thought. There were thousands of antibodies on the surfaces of B cells.

Once isolated, those antibodies could be used for study. For instance, if we knew what particular antibodies responded to particular pathogens, could we then figure out how the deadly diseases attacked or how the dance between self and alien took place?

Dr. Fauci told me the change led to a profound shift for immunology, turning practical a field that had been esoteric even as late as the 1970s and '80s. "All of a sudden, the immune system was having an impact on more diseases than you could possibly imagine," he said. He didn't mean that the immune system was having a new effect, but rather that it was now clear to scientists how powerful the effect was everywhere. "Cancer, autoimmunity, auto-deficiency, allergy."

These isolated and multiplied antibodies were known as monoclonal antibodies. They are changing your life, right now. Drugs built on monoclonal antibodies have become a dominant source of drugs in the early part of the twenty-first century. The annual market for these drugs is nearly $100 billion. They work by intensifying—or dulling, as the case may be—the performance of a particular antibody so that the body does a better job of attacking a life-threatening risk, like cancer, or, alternatively, dampening our elegant defenses so that the immune system doesn't behave so aggressively and cause autoimmunity.

The drugs have names like Humira and Remicade (which Linda and Merredith both tried in an attempt to try to slow their zealous immune systems) or ipilimumab, which has saved countless cancer patients, or nivolumab, which saved Jason. In the upcoming stories, you'll see the development and work of some of these miraculous medicines in an intimate way. In a general sense, the aim of these drugs is a relatively precise manipulation of the immune system, a molecular-level monkeying, rather than the scorched-earth tactic of previous drugs.

As a reminder, picture the difference between two cancer treatments, chemotherapy and immunotherapy. In traditional chemotherapy, toxins that destroyed fast-dividing cells got dumped into the body, ideally killing, say, a lung tumor, but taking out lots of healthy tissue as well. This was the proverbial war of attrition. The Festival of Life had to outlive the tumor *and* the treatment. With nivolumab or ipilimumab, as you'll see, the idea is to use molecular tinkering to unleash the immune system to attack cancer—using the body's natural defenses—rather than injecting bleach into the body and killing everything that moves.

This is complex stuff. Where are we in immunology's story?

For most of human history, infection, even modest infection, killed people with the terrifying regularity of an open wound, the ingestion of undercooked meat, the casual exhale of flu inhaled by another, pneumonia passed from hand to hand and wiped on the nose. Then over the centuries, scientists took baby steps toward understanding these infections and dipped a toe into how our bodies fought back. These scientists came from all over the world, which is worth noting because it shows the powerful, essential value to our survival of cooperation across national boundaries and cultures.

We got a big break with vaccines and antibiotics. These helped keep us alive without our really understanding how the immune system worked. More or less blindly, we squirted medicines into our bodies; they sometimes worked and often didn't, and we frequently didn't know why, one way or the other. But we began to chip away at the details too, particularly in the middle of the nineteenth century.

The T cell came from the thymus and seemed to play a huge role in mounting a defense, but exactly how it did so wasn't clear.

Ditto with the B cell, which came from the bone marrow, played a huge role, and seemed to have essential interaction with the T cell.

A Japanese scientist (Tonegawa), who studied in San Diego, then made a discovery in Switzerland that explained immunology's big bang: Our DNA rearranges itself in utero and forms millions of antibodies capable of binding to—and attacking—a trillion different antigens.

An Australian vet (Doherty) worked with a transplanted Swiss scientist to figure out that the T cell distinguished alien from self.

Then came a Russian and a final major discovery that came surprisingly late in the story of our elegant defenses. There isn't just one immune system, but two.

20

A Second Immune System

How are we able to eat food without our bodies' attacking it as foreign? After all, a banana isn't human, nor is bread, let alone a Philly cheese steak (which may not even be food, with all due respect to Philadelphia natives). We swallow, the food travels down into the stomach and intestines where acid breaks it down, and then nutrients are leaked into the body—tiny alien pieces but of tremendous survival value. How do our bodies know the difference between merely foreign and truly dangerous? It was a question that immunologists thought they had answered with, for instance, the discovery of the relationship between antibodies and antigens, governed by detectors like MHC.

Even in the search for AIDS, the presumption was that the action was all about this "adaptive immune system" governed largely by T cells and B cells.

Science was wrong. To answer the banana or cheese steak question, science required another foundational piece of information. Once again, the key discovery came from an international village of scientists.

Ruslan Medzhitov was born in the Soviet republic of Uzbekistan in March 1966. Eighteen years later, in college, he was living the clichéd life of a dutiful, freedom-starved Communist citizen.

"Every fall we had to go to the cotton fields for a couple of months. This was compulsory. You'd get kicked out of college if you didn't do that. It was primitive conditions. One time I was 'caught' by our department chair for reading a textbook in the field."

It was a biochemistry text.

"He said, 'I'm going to take away your stipend.'"

That was the bad news. The worse news was war. In the second semester of his freshman year, Medzhitov was called to military service. His head was shaved and he went to a plaza, where the recruits were divided into platoons of thirty and the groups essentially chosen at random to determine which would go to Afghanistan, which the Soviet Union had invaded in 1979. "The two groups before me and two groups after went to Afghanistan," he told me. "Many didn't come back. The ones who did come back weren't normal."

As he looks back at the fateful war in Afghanistan, the hostility of the crumbling Communist regime to anything foreign now looks to him a bit like an autoimmune disease. "You're trying to destroy what you perceive as nonself, and you destroy a lot of self," he said. "It's sort of like autoimmunity," he added. "It's exactly what's happening in the Middle East."

Political and cultural defense systems run amok, hypersensitive, reacting without checks such that they can no longer tell what will spare and preserve them—what keeps them in homeostasis—and what will be their undoing at their own hands.

After Medzhitov's military service, he returned to college, interested broadly in the sciences, not particularly in immunology, and got what appeared to be a huge break. He was selected, after multiple interviews, to go to the United States to study. "It was an unbelievable miracle," he gushes.

"I couldn't believe my luck. There was just one last step." He got a phone call one day from a man who told Medzhitov he needed to go through an orientation and asked to meet the young scientist in a park. "In retrospect, I always think: How did I not know how fishy this sounded?"

The man whom he met wore a suit and a tie. He "looked very vague. When I try to remember, there is no face. There's everything else without a face."

They talked about this and that, and the man asked to meet again a few days later. The next time they saw each other, the official appealed to the student's patriotism, saying, "You want to help your country, right?" Medzhitov recalled. "I'm thinking to myself: 'Oh, shit.' That's when I realized he was from the KGB."

The man knew everything about Medzhitov—his grades, his love of basketball. But the man didn't overtly threaten. He just explained that Medzhitov would be asked to gather classified information in the United States and transmit it back home. He was going to be a receptor for the Soviet Union's overheated immune system. He would be a T cell, doing surveillance in the United States. "'We're going to teach you to sneak into buildings at night,'" he recalled being told. That part sounded a bit like James Bond. "That was exciting. Everything else about it stunk. I tried to explain my point. 'I want to study and not be a spy.'

"The very next morning, I got a phone call from the Office of International Affairs. They said, 'Your documents got lost. You're not going anywhere.'"

He had stayed true to himself. It had cost him, dearly.

Then came another stroke of luck, or if you prefer, one of those random moments, a veritable random mutation in time and space, that led to scientific evolution. The spark was set off thousands of miles away from Medzhitov, on the north shore of Long Island.

In 1989, Yale immunologist Dr. Charles Janeway Jr. gave a speech at a symposium in Cold Spring Harbor, New York. In the lecture, he audaciously proposed to illuminate "immunology's dirty little secret."

The secret he was referring to was that the immune system was built fundamentally—essentially exclusively—around the dominance of the T cell and the B cell. This was the adaptive immune system, and I won't belabor or repeat here its deeply rooted history in immunology.

But Dr. Janeway was troubled by a crucial question, one so simple that it had until then been overlooked. How did the T cells and the B cells know which cells to attack?

You might think, once again, at this point, that the question had already been answered. After all, antibodies and antigens had been discovered and their interactions had been widely studied. The dendritic cell was understood to present information to the T cell. The presumption was that T cells and B cells know what to attack because they recognize antigens. Remember these? They are markers on pathogens—tags.

Dr. Janeway was vexed by a question his students had asked him: Aren't there antigens on non-harmful foreign substances? What about the nutrients from a banana we eat? What about a bacteria we inhale that is innocuous? After all, there are billions of bacteria around us, and many are not deadly. Presumably, these cells or organisms have antigens. Our elegant defenses must be assessing them, and rather than attacking them, leaving them alone or even integrating them.

"What was known was how the immune system sees the antigen. What was not known is how it sees an infection. Antigen and infection are not the same thing," Medzhitov said as he explained the simple logic to me. He told me this story because Dr. Janeway passed away in 2003 of cancer. (His *New York Times* obituary noted he was "often referred to as the father of the understanding of innate immunity.")

At the Cold Spring Harbor symposium, Dr. Janeway proposed the idea that the T cells and the B cells recognize antigens, lots and lots of antigens, *but they don't on their own know which ones to attack.*

"They say: I got something, but I don't know what it is. Is it your own pancreas or a vicious virus?" Medzhitov explains. Is it nutrients from a digested banana or HIV? "They cannot see the nature of the antigen. It could be coming from our own cells, from food, something that came in contact with our skin. But not all of that is infectious or pathogenic."

The T cells and B cells, he says, "detect something with exquisite specificity, but at the cost of not knowing what it is."

Medzhitov borrows an analogy from Pavlov's dogs to describe the nature of the problem that Dr. Janeway identified. Pavlov understood that his dogs would immediately salivate if they smelled food. They didn't do anything if they heard the ring of a bell. Then Pavlov paired the sound of the ringing bell with the smell of the food. The dogs associated the bell with the food and salivated.

Dr. Janeway had discovered that our adaptive immune cells don't attack only if they hear the proverbial ring of the bell (the antigen); they need another signal.

When Dr. Janeway proposed this notion, "he was largely ignored," Medzhitov recalls. "People thought this is just another crazy idea."

It didn't help that Dr. Janeway offered no proof. What exactly was telling the T cells and the B cells that the antigen they had identified belonged to something that deserved annihilation? What told them to leave the good stuff alone?

In a generic sense, Dr. Janeway proposed the idea of a "co-stimulatory" signal. This would be an agent, a message of some kind—from someplace—that would inform the T cell or B cell what it was looking at.

Back in the former Soviet Union, Medzhitov was in a Moscow library reading various papers when, while pursuing another subject, he came across Dr. Janeway's theory. He had more than a passing interest in immunology by this time, and when he read this paper, it had the extraordinarily powerful impact on Medzhitov of crystallizing a question that had long vexed him about how the human body dealt with the outside world.

"Just completely by accident, I read his paper. I thought: This is it. This explains everything," Medzhitov says. Prior to this, he'd realized, immunology was fascinating, "but it was a collection of stuff with no logic behind it."

Medzhitov paid a full month's college stipend to make a copy of the paper so he could study and read it over and over. It was 1991, and he had become obsessed.

Medzhitov typed up a message to Dr. Janeway on a big floppy disk. It essentially said: I'm fascinated by your theory, and here are some implications.

"After a week, he sent a response. It was a really memorable moment. He started discussing the theory with me. I was a nobody student from Moscow, and he was a very famous scientist!"

The Soviet Union was imploding. Amid the "vacuum of laws" that followed the Soviet collapse, Medzhitov made his move, securing a fellowship in San Diego. By early 1994, he wound up in New Haven, working for the man he'd come to idolize.

The pair were determined to prove that the T cells and B cells don't go into action until they get two pieces of information. While they recognize an antigen (a foreign substance, be it food or a virus), this information is largely meaningless without a second piece of information, which is a co-stimulatory signal that says "kill."

Where did that second signal come from?

In seeking an answer, researchers in the 1990s were acquiring their own supertools, in the form of computing power and programs that allowed a much deeper analysis of the seemingly invisible, such as wider mapping of the immune system at a molecular level. Among the tests that Medzhitov now had at his disposal was the ability to identify segments of individual genes. He couldn't see the entirety of most genes because the human genome—the whole of its sequence—hadn't yet been mapped. But the technology allowed him to map portions of individual genes. Here's how Medzhitov puts it: if you imagine a gene as a person, you might be able to map the foot and then make some inferences about the leg. Bit by bit, you could build a genetic profile of the whole person.

Or a fly. It was a fly that led Medzhitov and Janeway to their breakthrough.

They'd been searching in the dark for a way to prove the existence of a co-stimulator, a signal to push the T cells and B cells into action. Then they heard a lecture related to a discovery made in the mid-1980s in fruit flies. The finding was that flies with a mutation of a certain gene couldn't control fungal infections. The gene was named Toll.

The first time I heard *Toll receptor*, I assumed it was some metaphorical term related to a booth on a highway. In actuality, it comes from German, and means amazing or wild or great. (According to one history, this was because a German scientist, upon grasping the results of the study, exclaimed, "Das war ja toll": *That was amazing.*) It often goes by the name *Toll-like receptor.*

Medzhitov and Janeway thought it sounded, if not amazing, at least promising. They figured this Toll-like receptor may be responsible for helping the adaptive immune system discern what to attack and what to leave alone. What if it helped explain why our bodies don't attack a banana or our own spleen? The Yale scientists started looking for fragments of DNA that would be the human analogue to the fly's Toll receptor.

First, they found the gene, or fragments of a gene that looked like the human version of the one in the fly. Then they did experiments to see if they could show that the gene was not just instrumental but essential in causing the T cell to act upon a pathogen. One night in February 1996, Medzhitov was checking the lab results on his computer. This is one of those experiments that is too technical to describe and, in its own way, not the stuff of Hollywood; first there were some mixtures, or assays, and then the data was crunched digitally and the results came over the computer.

But those results? Now that part *is* the stuff of Hollywood.

Medzhitov and Dr. Janeway had found the fundamental mechanism that allowed the body to determine if it was dealing with a pathogen, a bad guy such as a harmful virus or bacteria.

This was the discovery of what happens at first contact. The Toll-like receptor is as elemental a concept as in all of our survival and in the science of immunology, and it had taken years to uncover.

"It was Holy Grail at the time, the dream result, to find something that provided evidence for a hypothesis that at the time only two people cared about," Medzhitov says. "It was eight o'clock at night and it was well known that Dr. Janeway didn't like to be bothered at home. I couldn't even contain myself and wait until the next day. I called him and told him the result: 'I saw induction in the genes.' He knew what that meant."

The discovery became the basis for our understanding of the concept of a second kind of immunity. It is called innate immunity.

The innate system shows up, discovers a pathogen, and mounts an initial but generic attack, meaning the attack is not specific to the pathogen. It can hold off the evildoers but often cannot kill them completely. That requires specific attacks from a particular T cell or B cell armed with the receptor or antibody that matches the antigen on the surface or inside the bacteria or virus or parasite.

The innate system informs the adaptive system: *I need help. Bring the heavies.*

The innate immune system scans organisms for the presence of one of a handful of key identifying markers that are shared by viruses and bacteria. For instance, most bacteria have wiggly tails. Toll-like receptors scan for these. Or they look for a particular variety of large molecules—called lipopolysaccharides—that characterize a class of bacteria called gram-negative bacteria (such as *E. coli*); or they look for nucleic acids associated with viruses.

Compare now several scenarios, one in which you get bitten by the cat, another in which you ingest a banana. In the first scenario, the cat's saliva trickles into the wound on your hand, setting off the cascade of immune cells, carried through opened blood vessels, bringing redness and heat. Among the cells at the scene are macrophages and dendritic cells with Toll-like receptors on the surface. The receptors can instantly determine whether the foreign substance entering the body has the hallmarks of a major pathogen.

If a pathogen presents itself—say, a noxious bacteria—not only does the immune system unleash a first-line attack but also the dendritic cells, now aware of the pathogen, begin their journey to find the T cell and B cell necessary to provide a more specific defense.

By contrast, when you eat a banana, the food travels down into your stomach and intestine. The gut breaks down the food, and nutrients leak into the body. Those nutrients, by the time they are broken down, may look much like "self" and thus not attract attention from the immune system, or our elegant defenses may identify the scraps of nutrients as foreign but not see any of the hallmarks of a pathogen. They have been accepted into the body, permitted to survive in the Festival of Life.

The role of the Toll-like receptor represents a relationship between human beings and the outside world that is as ancient as our existence. It was cultivated through epochs of evolution so that the human genetic code has developed the ability to scan for the ancient markers shared by hundreds of thousands of pathogens.

In a 2002 paper, Dr. Janeway and Medzhitov described it like this:

> The innate immune system is a universal and ancient form of host defense against infection. These receptors evolved to recognize conserved products of microbial metabolism produced by microbial pathogens, but not by the host. Recognition of these molecular structures allows the immune system to distinguish infectious non-self from noninfectious self. Toll-like receptors play a major role in pathogen recognition and initiation of inflammatory and immune responses.
>
> Thus, microbial recognition by Toll-like receptors helps to direct adaptive immune responses to antigens derived from microbial pathogens.

To break the findings down further: We are born with primitive detection mechanisms that can discern not only what is alien but what is *pathogen*. As a first-line defense, the molecules of the innate immune system recognize a large class of pathogens and signal the T cells: *That thing you just identified as alien is bad—go kill it.*

With this discovery, the major pieces of immunology had been put into place. Much was still to be discovered. But immunology suddenly faced a crisis that crystallized much of the science into a very practical threat.

A plague was afoot.

The greatest modern challenge to immunology and the immune system happened in the 1980s. Or rather, that's when it became clear the apocalypse lurked. AIDS led to a turning point in the story of immunology. The study had been so much about the lab and the mice, about inscrutable language and piecemeal science. Then came this crucible.

So our story turns too, moving more and more out of the lab and into the clinic, into the lives of patients, and into a new era of research. While basic immunology continued, there was an exciting new emphasis on applying the decades of hard-earned knowledge to more practical things, like the interaction of the immune system with sleep, stress, allergy, cancer, or nutrition, and like poorly understood symptoms that were actually autoimmunity. Various medical specialties—heart, lung, muscular, skeletal, and on and on—began to put to work the tools and knowledge of the 1970s. In that respect, what followed was an expansion of immunology.

It was spawned by the scariest disease modern medicine had ever seen.

Part III

BOB

21

Sex Machine

Bob Hoff thought he contracted hepatitis on Halloween night of 1977. It went with the lifestyle, he figured. He'd had genital warts and syphilis and various other STDs.

As a closeted young man from Iowa, Bob viewed sex not just as a preference but as an expression of self. "I was extremely promiscuous," Bob said of that period in his life. "I've visited every single bathhouse in the United States."

There was the Library in Minneapolis, Man's Country in Chicago, the Ballpark in Kansas City, and the Arena in Denver, and others in St. Louis and San Diego. The 1970s were a coming-out party for the gay community, an awakening for many gay men. As Bob put it, "I wasn't the only one out there." They'd lived closeted and in fear for so long, and they let loose with abandon.

Bob, a senior government litigator, was traveling the country, and he jetted around having unprotected sex. His wife, a flight attendant, also traveled frequently, giving him ample opportunity to have fun at home too. One day in 1978, Bob was working out at his gym in Crystal City, Virginia, where many in the D.C. political community lived, and he met a guy named Ron Resio. Ron had a triple doctorate and worked at a Navy base in Virginia helping update the F-4 Phantom fighter jet. Not the construction end, but the design side, the genius end.

"He looked like Conan the Barbarian," Bob recalled. Long hair

Robert Hoff, 1973. (Courtesy of Robert Hoff)

and big muscles. The pair became friends, and one day while Bob's wife was gone, they went to Bob's house and had sex.

Ron, it turned out, wasn't just another friend with benefits.

What happened next was one of the most excruciating trials that the human immune system ever passed through. It's also a story of how a search for a cure drew from the tremendous discoveries science had made over the prior fifty years. The bedeviling search to stop AIDS would also eventually come to draw from the exquisite immune system of Robert Hoff.

22

GRID

In August 1980, at Denver General Hospital, medical student Mark Brunvand was sent up to the ninth floor, critical care, a regular rotation for a third-year student. Years later, Dr. Brunvand would be Jason's cancer doctor. Now, in med school, he was forming the philosophy that would guide him in his career. His world view, like that of other doctors and researchers at the time, was being formed by a strange new disease and the havoc it wreaked.

On that August day, on the ninth floor, Dr. Brunvand went into the room of a patient who had come in with an unidentified illness. The man now lay in bed, hooked to a ventilator, unable to talk. Dr. Brunvand felt the man was trying to communicate through sad and terrified eyes.

Another med student told Mark: *Nice guy. We don't know what's going on. He's probably going to die. Looks like pneumonia, and he's gay.*

In a way, this is part of medical training; students take the labs and babysit the terminal patients. The labs in this case didn't make sense.

"Everybody was baffled. Nothing cultured out," Dr. Brunvand reflected. Then it looked like a parasite. "But we didn't get any confirmation."

They looked for reasons. What was unusual about this guy? Nothing to explain this. "We don't know if this guy has been smoking crack, has been exposed to toxic gas or other guys in the neighborhood."

Dr. Brunvand remembered looking at the guy and feeling completely helpless.

It was the kind of story unfolding all around the country.

June 5, 1981. The CDC put out case studies of five patients in Los Angeles. They were treated for pneumocystis carinii pneumonia. Two died. All were labeled "active homosexual." A lab at UCLA reported the cases. It's a novel place, this lab; it has been set up to combine clinical work with immunology. The researchers at UCLA discovered that the patients had "profoundly depressed numbers of T lymphocytes." T cells.

On July 3, a second CDC report came out reporting twenty-six cases in Los Angeles, New York City, and San Francisco.

Here's a snapshot of the kind of patient who showed up, baffling doctors.

At a bedside in Memorial Sloan Kettering Cancer Center in New York City, that same month, July 1981, a greenhorn physician, Dr. Mike McCune, looked at the racked body of a twenty-four-year-old man whose symptoms made no sense.

"His lungs were concrete," Dr. McCune said, reflecting back.

He'd been moved from Cornell, where they couldn't find a cause. He was surviving thanks to what McCune called a "super-duper ventilator" that managed to get air into his failing lungs. The patient was African American and had a history of intravenous drug use. In medicine the term *differential diagnosis* basically means: What's the likeliest cause among a list of probable causes?

"Cancer, cancer, cancer. What else could be causing it? An infection? But what kind of infection?" Dr. McCune said. "We put a tube down his throat and brought stuff up, and looked under a microscope. And what did we see?

"Not cancer. Not bacteria."

It was a parasite called pneumocystis carinii. Under the microscope, this looks like round clumps. The lungs of McCune's patient were swarming with these things.

The thing is, ordinarily they're not that dangerous. "You probably have them growing in you right now," McCune told me. "But your immune system is keeping them down."

Dr. McCune was transfixed. "I went back to the lab and thought: What's this guy got?"

The man held on for weeks, and then he died.

They were all dying.

Not Bob Hoff.

Bob's phone rang in mid-1982. The caller was Michael Ward. He was a good friend of Bob's and an undertaker at Fort Lincoln Cemetery. He'd been a lover of Ron Resio, a man Bob had had sex with too. Michael was calling with bad news and with a request. The news was that Ron had been admitted to Building 10 at the National Institutes of Health with an unusual illness. The request was that the NIH wanted to take the blood of Bob and four other men whom Ron had been with.

By now, there was a term for a new kind of STD showing up in the gay community. The illness was called GRID, gay-related immunodeficiency. Bob Hoff read about it in *The Blade*, a newspaper for and about the gay community in Washington, D.C.

The five men showed up at the NIH. They were met by a team

heading up a small, elite research group that had been set up by Dr. Tony Fauci. Team members included two accomplished physician-scientists, Dr. Cliff Lane and Dr. Henry Masur. Dr. Fauci was baffled, concerned—and fascinated.

"I looked at this and said, 'Oh my God, I don't have any idea what's causing this,' but when we look at the immune system, it's completely messed up. It's a disaster," Dr. Fauci said.

These guys started showing up who couldn't fight off basic infections, the kinds of viruses and parasites that the rest of us kick as a matter of course. The human defense system had been breached.

"Holy shit, if there ever was a disease I should be studying, this was it. This has to be an infection, but I didn't know what it was," Dr. Fauci said. "It's clearly attacking the immune system. It's an unbelievable situation where a virus is attacking the immune system. We've never had that before. We didn't know what the hell we were dealing with.

"I stopped everything else I was doing."

Fauci had found his dragon. Or was it really a windmill? Could it be fought or was it so elusive as to be practically illusory?

When Bob Hoff and four other men showed up at the request of the NIH to see their good friend Ron Resio, they were first asked to give blood in the auditorium at Building 10. Bob Hoff's blood test was a near disaster. The doctor, in search of a vein, nicked an artery.

"Blood spewed all over this doctor," Bob recalled. "He was scared to death."

The blood draw was something of a shot in the dark. Fauci and his team didn't know what they were looking for, maybe something in the blood, anything that might tell them what they were dealing with. At the very least, Dr. Fauci said, "we wanted to store their blood for future study."

After the blood draw, the men went to see Ron in the critical care ward. This once long-haired behemoth looked emaciated; he was covered with purple lesions, and tubes protruded from all over his body. Beyond the mystery illness, Ron was very interesting to Dr. Fauci because Ron had a twin brother. Might his twin shed light on what the hell was going on inside Ron's immune system?

That day, Ron's friends and lovers stared at him, shocked. They tried not to cry, because as Bob put it: "That would've made it about us, when this was about him."

After the men left the room, they let their emotions go. "Then we went to Glenn's house and we all had sex," Bob said.

You read that correctly. The group of men, having seen the first of their friends dying of something terrible, went to one of their homes and had an orgy.

Bob put it to me just as plainly as I've shared it with you. I asked what prompted such a response, and he said, "Well, we practiced safe sex." But there was more to his answer than that, and it was another instructive moment for me in the conversation about how we define ourselves in terms of self and other, just as the immune system seeks to do. Bob and his friends had one another, and they had sex as a defining characteristic and a sign they were not as alien as they'd felt growing up.

Plus, Bob told me, many of these men were part of Washington, D.C.'s elite in-crowd. One of them in the orgy that day had been a campaign manager for a presidential candidate. Many others in that inner circle—just not there that day—were part of the "upper echelon of the Republican Party," Bob told me. He'd been a Republican too for many years. They hewed to what let them feel safe and as if they belonged, one another and sex.

That was how the day ended for Bob, in catharsis. "For me, it was the last time I'd see a lot of those guys."

On September 24, 1982, the Centers for Disease Control and Prevention put out a report saying it had received 593 cases of what is now called acquired immune deficiency syndrome (AIDS). The condition that Dr. Brunvand had seen in Denver and Dr. McCune had observed in New York now had a name. Of the reported cases, 41 percent had died. Many of them had the parasite pneumocystis carinii; others had Kaposi's sarcoma or another opportunistic disease that was ultimately proven to have a viral basis. These were viruses that took advantage of a suppressed immune system. In many of us, such infections would be held in check and certainly wouldn't kill us.

There is a telling sentence in the CDC note: "The CDC defines a case of AIDS as a disease, at least moderately predictive of a defect in cell-mediated immunity, occurring in a person with no known cause for diminished resistance to that disease."

To repeat: *occurring in a person with no known cause for diminished resistance to that disease.*

Less than 1,000 cases had been reported. Still, the medical community took notice. The immune systems of these patients were so befuddled that they were failing to hold in check viruses and other pathologies that ordinarily caused no problem. And not just one pathology, but multiple ones. In other words, some new thing was unraveling our most basic and elegant defenses.

It is not an understatement to say that some big thinkers saw

Fourteenth-century Florence in the grip of the Black Death. (Wellcome Collection)

this as an end-of-days scenario. "We were in a full-blown panic. It was the plague," one immunologist told me. "We thought everyone would die."

At this point, I'd like to take a moment to show proper respect for some other plagues.

The 1918 flu pandemic killed up to 50 million people worldwide, according to the Centers for Disease Control, nearly 700,000 in

Bubonic plague in the lab. (Pete Seidel)

the United States. The CDC says that it is still not totally clear what made this flu so deadly. It has been hard to study in part because it is deadly to even deal with. But one important theory is that what made it so deadly is that the flu virus in humans—to which we'd adapted some immunity—had combined with a genetic variant from birds. What this meant was that many human beings did not have an antibody to combat the flu, even among the massive pantheon of antibodies we all get born with. This is what the CDC says: "Influenza experts believe that a pandemic is most likely to be caused by an influenza subtype to which there is little, or no, preexisting immunity in the human population. There is evidence that some residual immunity to the 1918 virus, or a similar virus, is present in at least a portion of the human population."

But not everybody died. That's because some people could mount an immunity. Some people did have the right antibody somewhere in their infinity machine. All hail the value of diversity!

Another big daddy of a plague was the Black Death, a killer of millions of people, including, at one point, as much as half the world's population in the fourteenth century. The *Smithsonian* magazine describes three different ways the plague attacks: through the skin, attacking lymph nodes (bubonic); through the blood; through the lungs. The deadly nature of the plague owed to several mutations in the bacteria that made it elusive to the immune system and easy to transfer. Our immune system, in the case of the lung version, was virtually helpless.

And a quick but important word about the bird flu that scared the living daylights out of infectious disease specialists in 1997. A three-year-old boy died in Hong Kong, and then seventeen more people died, struck down by a terrible virus found in birds. The idea was completely heretical when Dr. Keiji Fukuda, an influenza specialist from the Centers for Disease Control, landed in Hong Kong to do the forensics, but it turned out to be true—and all the live birds in the local markets were killed to avoid further contamination.

There is a key aspect to that flu that is consistent with other deadly viruses. The people who died weren't overcome by the flu itself but by their immune system's *response* to the flu. The immune system went into hyperdrive to stop what it perceived as an extraordinary foe. Massive inflammation followed.

"It was a cytokine storm," Dr. Fukuda said. "People were dying from having an overwhelming response."

But by the early 1980s, we'd seen flu before. GRID, or AIDS, or whatever the hell it was called, was something new. If you prefer your cup half full, there was a bit of good news. This potential pandemic happened as science had begun to get a handle on the immune system.

A massive machinery had started whirring that would change everything about how cancer was treated. It was all because of AIDS.

"AIDS was the 9/11 of immunology," a developmental biologist told me. "We suddenly got this panic and everybody started throwing money at immunology."

23

The Phone Call

Ron Resio, former multiple PhD muscleman, died of AIDS in 1984. Bob Hoff, his friend and onetime sex partner, attended the service, a Navy funeral because Ron had served his country. It was the first funeral of someone who died of AIDS that Bob would attend. Later, Bob reached a point where he couldn't attend another friend's funeral; by then, he had attended dozens.

In the D.C. area, "five or six guys were dying a week," Bob recalled. "People would disappear on a daily basis. It was an onslaught."

In 1984, 3,454 people died of AIDS. It was going to get much worse. More than four times that amount would die four years later, and then the disease exploded globally.

Regularly, Bob said, American gay men would die, and their parents, disavowing their sons' sexuality, would disown the surviving partner, wouldn't invite him to the funeral, or would clean out the house and not give the partner his things back. The parents had decided they were self, and the surviving lover was other, an alien in their midst, and that their son had been an alien too, estranged even in death.

This was exactly how members of the prominent gay community in Washington, D.C., felt they were suddenly being treated by President Ronald Reagan. Bob knew all the operatives, who in turn knew Reagan and Nancy, the first lady, and the operatives

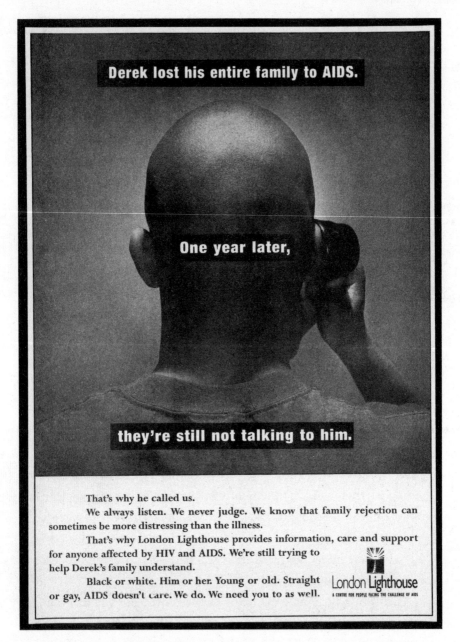

AIDS killed and society turned on gay men as if they were nonself, alien. (Wellcome Collection)

knew that Reagan liked them; some speculated his son was gay. "We couldn't believe he flipped on us," Bob said. Reagan's administration was widely criticized for its slow response to the AIDS crisis. This turned Bob, a lifelong Republican from Iowa, into a Democrat. Gay men were sick, and other people treated them as if they were toxic.

The community banded together. Bob, a lawyer by trade, had a real estate license. He tried to get gay men to buy property before they became sick "because money speaks." He wanted them to have some power, a voice. He and his lover at the time had a house on Fire Island, a gay mecca outside New York City that was the setting for weekly dinner parties and sometimes a refuge. At one point a friend who was in the Air Force got booted out for his sexuality and his condition, AIDS, and he showed up at Bob's house on Fire Island. That night Bob was upstairs in the house when he heard a thump, but he didn't pay too much attention to the sound at the time. The airman, too distraught to go on, had mainlined cocaine, "committed suicide in my living room."

One day in 1984, he recalled seeing a guy named Bill who had been "the most beautiful man I'd ever seen." Now Bill weighed 95 pounds and was a walking purple lesion. Death was everywhere and inevitable.

There really was no treatment, nothing that could be done. The scientists at NIH, led by Dr. Fauci, had thought maybe they could use the bone marrow—the source of immune system cells—of Ron Resio's twin brother to bolster Ron's immune system. The idea was that they'd take out Ron's marrow, which seemed unable to handle the virus, and replace it with healthy marrow that matched Ron's own. No such luck. "The virus destroyed the transplanted marrow," Fauci said. It brought certain death.

In late May of 1984, Bob went in for a regular physical. His doctor saw evidence of an irregular heartbeat and sent him for a follow-up. False alarm. He got the news in a phone call at his government office on June 8.

"Bob, I have some good news and some bad news. The good news is, we got a misreading on the heart test. The bad news is, you're HIV-positive."

Just like that.

"It didn't come as a surprise," Bob said matter-of-factly as he recalled the moment. "I was as exposed as anybody. I realized I was not going to get through it. I had a year or two to live. It was a death sentence, and there was nothing I could do.

"I realized I was just like everybody else."

He was most certainly not.

24

CD4 and CD8

At the highest level, there are two ways to understand and stop the spread of a pathogen like a flu virus or HIV. One path is to examine the chemistry, the biology, and the response of the immune system—the antibodies, the hard science. The other is to look at the circumstances surrounding a disease or outbreak, the epidemiology. What behaviors and broader factors seem associated with the disease? Does it happen in poorer areas with filthy water, or where certain foods are consumed, or where the air quality has changed?

Is it associated with sex?

In the first few months of the AIDS outbreak in 1981, the epidemiology said: "This is happening among clusters of men having very, very active sexual contact and, very soon thereafter, we found it was in injection drug users," in the words of Dr. Fauci.

This limited information still had great value to immunologists. It said that they were likely dealing with a virus. This was so for two reasons: It was transmissible like a virus, and importantly, it was unlike a bacteria or a parasite—in part because those can typically be seen in the tissue. Remember that a virus hides in cells. That can make it very hard to detect, even with sophisticated tests. The virus does slip out of cells as it goes from one to the next, but if you don't know what you're hunting for, "it's like looking for a needle in a haystack," as Dr. Fauci said.

Then at the end of 1984, there was a remarkable meeting at the CDC in Atlanta attended by some of the brightest minds in medicine. One attendee, Jack Dunne, described the mood: "Everybody is like: What the fuck? Nobody gets this. It was pre-terror."

One woman got up in front of the packed lecture hall, with perhaps a thousand people in attendance, and offered head-twisting epidemiology. She put up a graph with two axes. On the y-axis was the severity of the disease, and on the x-axis was "the number of fisting [anal fist-fucking] events per week."

The intimation was, there was something involving the ripping of tissue.

"My own hypothesis was that the people who were sickest used amyl nitrate," Dunne said. Amyl nitrate, known colloquially as poppers, relaxes muscles like those in the anal cavity to allow easier sex. Bob Hoff and his cohort used them all the time.

"Everybody was trying to figure out the mechanism of action."

Meanwhile, there was, of course, one more big-picture data point: Everybody was dying. "I refer to them as the dark years. It was terrible, horrible. This was an inexorable process," Dr. Fauci said.

How did this infiltrate the body and befuddle the immune system?

On the hard sciences side, a clue had showed up with the very first patients.

The 1970s, that decade of explosive learning about the immune system, had yielded important clues about the depth and subtlety of the immune system. One such clue is that the T cell itself is much more complex and multifaceted than had been previously understood. In fact, it became clear in the 1970s that there were fundamentally different kinds of T cells, core immune cell soldiers and generals.

"To that point, one T cell looked the same as another T cell under a microscope," Dr. Fauci said.

The two main kinds of T cells that had been discovered were known with characteristic blandness as CD4 and CD8. CD4 T cells are called helper cells, and they induce action by other immune system cells; CD8 cells are killers. They do the dirty work. Or, if you prefer, CD4 cells are the generals and CD8 cells are the soldiers.

The initial tests suggested to researchers that men infected with this syndrome had sharply lower counts of CD4. Given that relatively little was known about the immune system, it was goddamn lucky that one of the things known was implicated.

"It was curious that their CD4 cells were way down, and some of them even had an increase in CD8," Dr. Fauci pointed out.

It looked a bit like the immune systems of the ailing had very few generals left.

There was another lucky break. It had to with a discovery made a few years earlier that, on its face, had nothing at all to do with AIDS or T cells. It concerned cancer.

In 1965, a physician and pioneering scientist named Robert Gallo arrived at the NIH and started treating children with acute leukemia. "Mostly unsuccessfully," he wrote in a journal history. It was rough, dealing with terminal cases—"a vivid experience and one which made me absolute in a decision to be fully involved in laboratory research and not return to clinical medicine."

In the course of studying leukemia, Dr. Gallo started looking at retroviruses in animals. These viruses were known to cause leukemia in some animals. That's why he was studying them. It wasn't known if there was such a thing as a human retrovirus. Dr. Gallo wrote that looking for one was "an unpopular goal at this time, considering the decades of attempts and failures." The effort to fight cancer had been down this road before; plus, there was "little evidence" of leukemia-causing viruses in primates.

Finally, retroviruses were generally easy to identify in animals, so if there had been one in humans, shouldn't it be more self-evident?

What is a retrovirus? A nasty little bastard, typically more cunning than your usual viral fare.

Understanding the retrovirus requires the slightest explanation of basic genetics. DNA is the biological master plan. It dictates an organism's characteristics and traits. RNA helps execute the plan. I think of DNA as the architectural blueprint, and the RNA is the general contractor. RNA puts the plan into action, and it instructs lots of "subcontractors," like cells and proteins.

A retrovirus adds a new and unexpected twist.

In a retrovirus, the RNA turns viral; it has contracted a virus. The viral RNA is equipped with a special enzyme that causes a process called reverse transcriptase, which turns the RNA into DNA. In other words, the virus causes the process to go in the opposite, or reverse, direction from the typical genetic process by which DNA instructs RNA. Here RNA has become DNA, and that DNA integrates into the nucleus of the cell and into an organism's own DNA. Thus this virus has essentially co-opted the organism to make copies of itself—copies that are hard to detect. It squirts out of the cell as viral RNA, infects another cell, and the cycle goes on.

This was generally understood when Dr. Gallo entered the picture. He was the first to discover a retrovirus in human beings. It was called human T-lymphotropic virus type I. HTLV. This is a retrovirus that infects T cells. We understand much more about it now than was grasped then. We know now that the virus is inside a measurable portion of the population, up to 1 percent in some regions of the world, according to the National Centre for Human Retrovirology in London. The organization notes that most people have the virus inside them for years and do not suffer disease from it. Somehow the immune system keeps it in sufficient check; only one person in twenty develops disease.

One such disease is adult leukemia. This is what Dr. Gallo was looking for, and he found it, a link to cancer. He also discovered an important marker of the retrovirus that explains in part why I'm telling the story. Sufferers had a low CD4 count.

The earliest researchers of this deadly plague, not yet even called AIDS, had their first clue. It had a characteristic shared by a recent discovery of a human retrovirus. "People argued: It attacks CD4 positive T cells. Something is killing them. Maybe it's another form of retrovirus," Dr. Fauci said.

HIV is the virus that causes AIDS. The story of the discovery of this core connection has been told many times and very well, and I won't repeat it here in a lesser fashion. In the utmost shorthand, it was discovered through key work by Dr. Gallo and Luc Montagnier and Françoise Barré-Sinoussi in France, and many others less celebrated who made key contributions. (There was a dispute over exactly who deserves the credit and whether Dr. Gallo's work was complicated and overlooked by the Nobel Prize committee, but that's a discussion for a different book.)

What is relevant here is that a collection of great scientists figured it out, and their work was built on the incredible significance

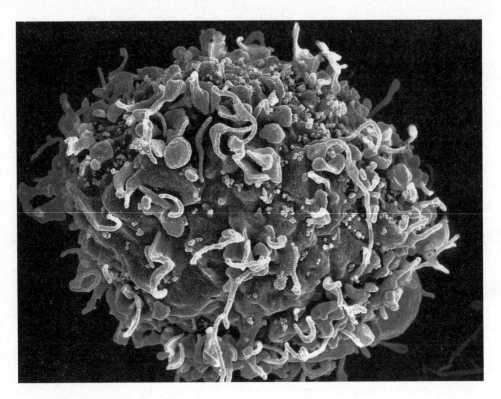

Small buds of HIV infecting a human immune cell. (NIH)

of Dr. Gallo's discovery of HTLV—"the sine qua non," Dr. Fauci said. "Without it we would've been nowhere with HIV.

"What happened next happened very quickly."

A test was developed to screen for the disease. You'd think that would have been a positive development, but initially at least the news was not just bad but terrifying. For years, Dr. Fauci said, the men coming to the clinic for treatment were in the final stages of their lives, but their numbers were relatively small, suggesting HIV was somewhat contained. But when scientists and doctors started administering the test to seemingly healthy people, they discovered that the infection was widespread.

"To our amazement and horror, we found out the sick people were just the tip of the iceberg. Thousands and thousands and thousands of gay men who were not sick tested positive for the antibody," Dr. Fauci said.

There were 16,908 deaths related to AIDS in 1987, according to the *New York Times*, 20,786 in 1988, 27,409 in 1989, and 31,120 in 1990.

The sufferers of HIV and AIDS were society's throwaways.

There was at least one extraordinary exception. Magic.

25

Magic

On November 7, 1991, Stephen Migueles, a medical student at the University of Miami, was ironing a shirt and watching the news when one of the world's greatest athletes, Earvin "Magic" Johnson Jr., appeared on television for a special announcement. Magic wore a dark suit, a white shirt, and a gray tie with a hint of red.

"Because of the HIV virus that I have attained, I will have to retire from the Lakers."

The eventual *Dr.* Migueles was riveted, like so many, but perhaps his interest went deeper. Dr. Migueles worked in the AIDS ward trying to stop one of the deadliest viruses ever with the equivalent of Band-Aids.

Dr. Migueles brought extra baggage. He was coming out of the closet. No easy task for him, having grown on up in a Hispanic, deeply Catholic family.

"I knew what I was, but I hadn't blossomed into my full self, if you will," he told me. He'd come out to his family and it hadn't gone well. At the time, his parents were devastated.

Then he was watching the men in the ward die. "I was trying to be true to myself, and I saw people around me who had come out and were proud of who they are and dying because of it. It was a scary crossroads."

Magic Johnson's revelation meant something to Dr. Migueles. "He was more mainstream," Dr. Migueles said, but that wasn't all.

"Most people you learned about who had AIDS were celebrities and seen as dying. Magic seemed to be a little different, seemingly robust. He looked like he was doing great."

He was lucky, of course. Only days after Magic's announcement, Freddie Mercury, the operatic rocking lead singer of Queen, announced he had AIDS. He died on November 24, 1991.

Four years after Magic's revelation, the Food and Drug Administration approved a drug called saquinavir. This was the first protease inhibitor.

Protease is the enzyme in HIV that helps the virus mature once it leaves the nucleus of the cell it has infected. If the enzyme gets inhibited, the virus doesn't mature. The virus doesn't spread. The immune system remains intact. The patient doesn't die.

"This is some of the most hopeful news in years for people living with AIDS," said Donna Shalala, then secretary of health and human services, a federal government cabinet position.

The inhibitor was part of a broad strategy that had been emerging aimed at defeating HIV by hitting it at various points in its "life cycle." For instance, the first major drug had been azidothymidine, or AZT, which was approved in 1987. AZT interferes with the enzyme that causes the retrovirus to transform from RNA to DNA.

On its own, AZT had some effectiveness and some side effects. It also could lead to a drop in neutrophils, those critical immune system cells. It could cause anemia, which is a drop in the red blood cells that carry oxygen.

Together, AZT and a protease inhibitor led to a significant increase in CD4 cell counts. (If you want to geek out, values of CD4 cells rose by 30 or 40 cells per milliliter of blood, a significant figure when the amount in a healthy person is 800 cells per milliliter of blood. Better yet, the CD4 count didn't drop.)

It was a turning point in the battle against HIV.

By 1997, the death rate due to AIDS had dropped 47 percent. AIDS fell out of the top ten causes of death in the United States, plummeting from eighth to fourteenth.

But it wasn't the answer to what was happening with HIV. Rather, the drug was like a somewhat effective antibiotic or vaccine. It didn't explain why some people seemed to be able to fight it themselves. This deadly disease left some people untouched.

A key insight came from Patient 1.

This man was a hemophiliac, meaning his blood didn't clot. Bad news, of course—when you can't clot, bleeding is prolonged, even indefinitely, and you can die without treatment. To counter this rare genetic condition, the man had received regular infusions of the protein that helps blood clot. One of his infusions was contaminated with HIV, long before it could be tested for.

"Patient 1," said Dr. Mark Connors, a Philadelphia native who had come to the NIH after medical school and pediatric training and fallen in love with pure research. An NIH colleague came to him in 1994 and said, "Dr. Connors, we've got this highly unusual patient."

The hemophiliac was in his twenties, and he had HIV but no viral load, the term for how much of the virus coursed around inside a person. With HIV, the viral load typically took a fascinating path. Initially, it would spike so that there were a million copies of the virus in each milliliter of plasma. (One patient was studied whose load spiked to 5 million copies.) Huge numbers. Then, however, the viral load would typically fall sharply during a chronic phase of the illness and then spike again as death neared.

The hemophiliac had little viral load. The guy wasn't sick.

With the benefit of hindsight, this might look inherently interesting, but Dr. Connors and others weren't so sure. There could've been a number of factors, including the simple possibility that the man had gotten a weak version of the virus.

Dr. Connors was put in charge of figuring it out.

Enter the mice.

The researchers at the NIH did a nifty trick by injecting the hemophiliac's cells into an immune-deficient mouse. They stripped out the mouse's immune system. The reason for this is that, as you now know, if the mouse had an immune system, it would have rejected the human cells as foreign. Now they had a mouse infected with replicating versions of Patient 1's cells—all kinds of cells, white cells, red cells, other cells.

The mouse didn't reject the human cells, creating a kind of living laboratory. Lo and behold, the mouse didn't get HIV. Again, this seemed important but raised the possibility that the hemophiliac's version of HIV was weak, not necessarily that the hemophiliac's cells were fighting the illness. Incidentally, as you might've inferred, the mouse ultimately died a horrible death because the human cells reacted against the mouse cells, so-called graft-versus-host disease.

Then came the Bingo Experiment. They gave mice the hemophiliac's cells, but this time they tinkered with the T cells. They did so by giving the mouse an antibody—that highly specific protein involved in detection and defense—that would pick up and attack the CD8 T cells of the hemophiliac. In other words, the mouse wouldn't reject all of the foreign cells, just a little piece, a key section of the T cell.

This time, the mouse contracted HIV. That pretty much nailed it. This was, is, a CD8-dictated mechanism. Bingo. HIV would win, unless the body's T cell foot soldiers unleashed an immediate effective response.

Subsequent studies in monkeys reinforced the discovery. The studies showed that the primate immune system, when artificially depleted of CD8 cells, lost control of the virus.

Bob Hoff, and a handful like him, helped tie the evidence together.

26

The Prime

In March 1998, Dr. Migueles, the young AIDS investigator from Miami, had finished his medical training and began a round of interviews to see where he'd go next. He knew he wanted to continue to work on HIV. He had lots of opportunities. But only one possible miracle awaited him. He discovered it—where else?—on the eleventh floor of Building 10 at the National Institutes of Health. So much great research had been done here, by Dr. Fauci and Dr. Dinarello and others, not just on HIV, but on the basic science of the immune system and its connection to a myriad of diseases.

Now Dr. Migueles came to interview for a fellowship. He met that March day with Dr. Connors in a small office that, coincidentally, Dr. Connors had inherited from Dr. Fauci. During the interview, Dr. Connors told Dr. Migueles that he and his team had started looking at a small group of HIV patients who hadn't seemed to be getting sick.

The interview took an enthusiastic turn. "That's unbelievable. That's got to be where the answer lies," Dr. Migueles said.

"I know, right? Isn't that amazing?"

Dr. Migueles told Dr. Connors about a patient he'd taken care of at Georgetown whose symptoms just didn't add up. "This woman comes in, she's incredibly sick for six days. Then she's fine. I was like, am I losing my mind?"

Dr. Migueles suspected the woman belonged to some curious, if not revelatory, group of HIV patients who were defying everything known about the disease. But he didn't have a name or context for what that might be. Dr. Connors had collected a handful of these people and started testing their blood. Were they just having a delayed onset of symptoms, or was something else going on?

Dr. Migueles was offered the job and took it. He wanted to work with Dr. Connors to cure HIV.

At that time, the so-called AIDS cocktail was having an impact on the death rate. That was relatively good news, particularly in the United States, where, as I mentioned, AIDS had dropped out of the top ten causes of death.

Still, every minute of 1998, an estimated 11 men, women, and children got HIV. Overall, 5.8 million people worldwide were newly diagnosed with AIDS, bringing the total of people living with the disease to 33.4 million, according to UNAIDS, a United Nations organization cooperating with the World Health Organization. Deaths globally in 1998 were 2.5 million, the most in any year, and the total dead from the epidemic was just shy of 14 million. The disease continued to be focused in developed nations but was increasingly spreading to emerging countries, with 70 percent of the people infected that year in sub-Saharan Africa, UNAIDS reported.

"The epidemic has not been overcome anywhere," the report reads. "Virtually every country in the world has seen new infections in 1998 and the epidemic is frankly out of control in many places."

And even where science and medicine had made great strides,

with the cocktail, there were powerful side effects. The drugs increased patients' vulnerability to diabetes, for instance. Perhaps this was not surprising, given the delicate balance of the immune system; strengthening it to fight HIV meant triggering echoes that, in this case, seemed to cause the body to attack itself and its ability to process sugars. Yes, it beat dying, but there was also nothing fun in developing a "buffalo hump," which was the nickname given to a condition common with the cocktail that caused fat deposits to rearrange in the body, notably in the shoulders.

One HIV-positive man, Brian Baker, began to develop a buffalo hump. He'd been diagnosed in 1993 when he was thirty. He worked in a record store and as a disc jockey. He lost fat in his cheeks, the layers of skin on his lips fell off. His moods swung. He had to go off the meds for a while. He was alive at least.

Soon he'd meet Bob Hoff, and a romance would bloom. In the meantime, Bob felt like a cornered man, watching all his friends die, waiting for his own shoe to drop.

"I felt that at any point in time I was going to be dead," Bob reflected. This was his experience of the mid- to late 1990s: inspecting his body for purple spots and his mouth for white fungus. He couldn't make sense of what was happening, and his confusion was combined with mounting survivor's guilt. "I would meet people and it was just unbelievable, they all died. I'd make new friends and all those guys died." He stopped wanting to go out at all. He likened it to the way his dad's friends had died in World War II, and before that, to how his mother's friends had died of the Spanish flu.

"Pandemics come along and kill people and wars kill people, and this was my turn at the barrel," he said.

Why not him?

He had a theory as to why he was still alive. Maybe, he thought, it had to with a healthy diet and with a colonic cleansing routine he did regularly. He thought that maybe his immune system had gotten so distracted by this process that it couldn't be overtaken by HIV. It didn't make a lot of sense, but what did or could?

By now his blood had long since been collected by the NIH; recall that he'd gone there years earlier with dying friends. He wasn't yet marked for study, though. He was merely one of the people that the NIH kept an eye on, given that researchers didn't yet know if he was simply destined to get sick. He'd go in every six months and give some more blood. He kept on living, asymptomatic.

Then he got a call to come in to meet with Dr. Migueles.

When Dr. Migueles was first hired at the NIH, he joined a meeting with the other investigators trying to figure out what they hoped to learn from guys like Bob Hoff. Dr. Migueles was the junior guy in the room, and he made a list of all the possibilities that might explain the molecular mechanism that made immune system marvels of these mortal men. It was needle-in-the-haystack work.

Given all the complexities of the immune system, a plethora of possible pathways might be saving these men. Could it be that they had gotten a weakened strain of the disease? Could it be that they had immune systems trained previously through some particular set of circumstances of diseases, or that they had a peculiar way of binding to the disease or communicating about it to other parts of the immune system?

Dr. Migueles made a long list of options, and the team set about trying to eliminate the irrelevant ones. They needed the vaccine or medicine that would bolster the immune system, and they were up against the clock. People were dying.

When he first met Bob, Dr. Migueles was working his way down that list of possibilities. It was December 10, 2007. Bob figured to offer further evidence.

"You have an immune system that is constantly fighting," Dr. Migueles told him. Bob was a "long-term non-progressor," in the language of the field. It should've been great news, at least to him personally, but Bob felt malaise. "There's no joy in being a survivor."

And, Bob recalled, he was told, "This is not a get-out-of-jail-free card." Bob was cautioned that he could still die if his immune system faced another assault—from, say, hepatitis, shingles—another debilitating attack that required his immune system's full attention.

Dr. Migueles said that he wanted to start studying Bob's blood to try to look for markers that might help explain Bob's own survival and help lead to a cure, a real cure. Of course Bob agreed.

At the time, Dr. Migueles told Bob that he had a theory that Bob had better-responding CD8 T cells than other people. He told Bob: "Your immune cells respond more vigorously to the virus than the cells of other people."

But this alone was essentially unsatisfying to Dr. Migueles and other researchers. To find a cure, to dismantle AIDS, they needed to know not only *what* the immune system did, they needed to know *how it did it*.

In the late 1990s, Dr. Migueles and fellow researchers at NIH—along with other researchers around the world—found a major clue that distinguished Bob and others like him.

Many so-called elite controllers, patients like Bob who keep HIV at bay, have a gene that impacts the way the immune system recognizes foreign invaders. Specifically, they share a genetic variant called HLA-B57. HLA stands for human leukocyte antigen. That's the human version of MHC that Dr. Doherty and others discovered years earlier and for which they won a Nobel Prize. The HLA is essential in helping the human immune system distinguish between things that are self and things that are foreign. In Bob and other elite controllers, this key gene, B57, seemed different. In the first study of elite controllers, eleven of the thirteen had this gene. By comparison, only 10 percent of the population as a whole have B57.

This was a very powerful discovery. It essentially identified one likely genetic underpinning of an immune system that could fight off this version of a plague—a key piece of DNA for unleashing an effective T cell response to HIV.

Further, Bob and the other elite controllers weren't surviving because their virus strain was weak. It was just as potent as strains killing left and right.

"They don't harbor wimpy viruses," Dr. Migueles said. He knew that they were seeing a powerful immune system variation. "This is evidence of what the human immune system is capable of doing. They are alive with infection we thought uniformly fatal but acting as if they have the herpes virus, and the virus is sitting there doing very little."

There was a third key discovery. It now appeared that Bob and the other elite controllers had survived likely due to a very specific moment in the interaction between their immune systems and HIV: the first point of contact.

"The evidence is pointing us to what we call the prime—the priming event. It's when the immune system first sees virus," Dr. Migueles said. "We suspect people like Bob start down the road to being an elite controller right at the beginning."

These are major revelations, particularly the idea that the way you deal with a disease might well be dictated by this idea of a prime, or first point of contact. The initial response, whether to flu or HIV or a cold, might well echo through the immune system. The right first response could save your life, not that you have particular control over such a thing. However, knowing this can inform the way we build medicines, or study individuals to see their susceptibility to various viruses, say, through genetic testing. Some of this is yet to be foretold by science—but is now within its grasp.

Indeed, the sum of the work done at the NIH has led to a much deeper understanding of the immune system. Such essential science "has relevance for inflammatory-based disease, autoimmunity, and cancer," Dr. Migueles said. The papers the scientists have written are seeds of medicines and treatments and, in particular, of vaccine development. The way that elite controllers react is based on a "common pathway" of how our elegant defenses work on a molecular level.

Dr. Migueles said that intensive study of HIV has helped develop "a flow chart of the multiplicity of relationships" in how the immune system cascade works. "That's where the treasure chest is."

Perhaps the biggest part of the canvas is how this research, along with lots of work from many places, led to the most important conclusion of all.

"People are no longer dying," Dr. Migueles said. The cocktail that saves lives, going all the way back to AZT, includes leaps in basic immunology, including those attained by the team at NIH. That work has had to keep going because HIV, like all organisms, continues to evolve, to survive, and to evade detection not just by the immune system but by the drugs.

"It's an arms race," Dr. Migueles maintains.

Another way to look at this arms race is from a social perspective. "This was a death sentence. People were terrified and nobody cared and Reagan wouldn't say the word," Dr. Migueles said. "Their own government had betrayed them. So they took it on themselves to be their own voice.

"This wouldn't have been done if they hadn't mobilized. It was miraculous."

They acted in their own defense, a social complement to their immune systems, calling out: We are not alien. We are part of society, we are self!

That notion has since led to many movements of medical self-empowerment, such as the crowds who walk for breast cancer and the sports figures who mobilize awareness around a particular disease.

In the end, key takeaways from Bob's story and key lessons about our collective health come from how we relate to each other on a social and political front. And Bob had his own happy ending. But before I bring his story and medical contributions to a close, I want to fill out the broader scientific picture by telling you about a different group of people, the ones whose immune systems are too powerful.

Part IV

LINDA AND MERREDITH

27

Linda

Linda Bowman came into the world in March 1960, the second-born in her family, and that status helped define her. Her big sister, Joanne, was two and a half years older. In the race of life, Joanne was Linda's rabbit—the thing to chase. If Joanne had homework, Linda wanted to do it. Linda excelled at math in particular; she was so good, she had skipped third grade. The larger truth was that Linda could apply herself, loved to do it, had that internal drive that only some have and most don't.

The first place she put that drive was into horse riding. When Linda was seven, her parents took her and her older sister to Wyoming to a family ranch, where you got to play cowgirl. Linda started playing for keeps. Back at home in a community north of San Francisco, Linda spent her afternoons and weekends practicing at a barn. Her family was privileged but not rich, her dad a midlevel executive at Chevron, and they lived in an Eichler home in Marin and got Linda her first quarter horse when she was ten.

She tried to stay lean so she'd look great on that horse. There was a period when she was around fourteen years old when Linda, of her own accord, would go on an all-protein diet for several weeks at a time, a regimen that was the precursor to the Atkins diet—meat and eggs, her only snack pork rinds, occasional cottage cheese. "My parents were a little worried, but I didn't have any eating disorder." She just liked to win. But horseback riding competitions were subjective. She hated not being in control of her results.

"It's what I love about golf." She started to attack the links as she had the stable.

At about this time, Linda first exhibited a health oddity. For years, her stomach had bothered her, even predating the periodic diets. Mostly it was constipation, sometimes terrible gas.

When she was fifteen, she went to play a round of golf with her parents at the Richmond Country Club. Just prior to tee-off, Linda went to the clubhouse restroom and had a bowel movement. It was partly a huge relief because she hadn't defecated in days. But immediately afterward she also felt weak and dizzy.

Her mother saw her, wobbly, heading from the bathroom and to the first tee box.

"What's going on?"

Linda explained, then swallowed some water and tried to shake the feeling.

Her mother responded: "Oh no. I hope you didn't get my stomach."

Linda's mother, Carol, suffered from irritable bowel syndrome. This is a condition that causes a range of stomach disorders—pain, constipation, diarrhea, gas. It is not an autoimmune disorder per se, but it can often involve inflammation, which is caused by an excessive or prolonged immune response. It is a cousin of irritable bowel disease and Crohn's disease, which are autoimmune disorders characterized by excessive inflammation. Imagine if the plumbing inside your body became inflamed, red and painful, swollen. For one thing, this causes physical discomfort simply because of the tight confines of your body; the space inside you has been engineered to near perfection by evolution, no space wasted. So when things swell, it hurts—potentially a lot.

Linda went right along, gifted, sure, but also willing and forging one success after the next. She worked her way into a golf scholarship at Stanford, eventually graduating with a degree in economics. Then she was tabbed to play on the European golf tour, at the time a struggling operation. The American women selected for the team, in addition to their golfing ability, all looked good, which was part of the marketing effort to sell the sport. This led to fun years for Linda, 1982 to 1985, before she'd had enough and moved on to the next stage of her life, as an MBA student at Stanford.

Linda married a man who would become a partner at one of Silicon Valley's big-name law firms and took his last name, becoming Linda Segre. She joined the Boston Consulting Group, an elite organization of consultants where she headed down the partner track, matching her husband work hour for work hour. She'd call him from her office at eight P.M.

"How you doing?" he'd ask.

"I could use another hour."

"Me too."

He'd swing by to pick her up in the Porsche 911 at ten.

With success came added responsibility and pressure. She met each challenge. That's how she saw it anyhow. One time, in 1989, she was vying for a project and stayed up ten nights in a row to compete for the deal. She won it.

"There were very, very few women, and plenty of very, very smart people and I felt a little insecure," she reflected. "I can prove I'm as smart as the rest of you guys. I did it by just killing myself."

Her husband worked no less hard, she recalled. Just like so many people in Silicon Valley—and in New York and Hong Kong and

London and lots of other type A enclaves. Many of those people do not get autoimmune disorders. So this lead-up is not intended to suggest that Linda brought her condition on herself. Her genetics were expressly at play too.

But it is fair to say that Linda was building a life that was not consistent with her own limits—nor with those of most people's. She was losing track of what was true and consistent with her, what was her real self. In a certain way, her life was being driven by the pathology of nonstop work, a foreign invasion that was threatening not just her emotional health but also her physical health.

In the late 1980s, the stomach pains got worse. Once every few months, she'd have such bad gas that she'd come home and crawl into bed. The swelling would be gone by morning. She kept pressing the limits until the bottom fell out.

In early September 1995, Linda gave birth to a son. He was the couple's second child; their daughter was two. The family lived in San Mateo, a comfortable suburb south of San Francisco. It happened to be just a ten-minute drive from one of Boston Consulting Group's major clients—a billion-dollar financial services company—and Linda was playing point on the account. The client relied heavily on her.

Linda convinced herself that she could continue to have it all. She took ten days of maternity leave. She was exhausted. "I would be taking calls at midnight and be up with my son every two hours breastfeeding."

She got a terrible sore throat that December, as bad as she'd ever had. She suspected it was strep, a highly contagious illness caused by the streptococcus bacteria. Typically, it is treated by an antibiotic. Not in her case. "I didn't have time to go to the doctor."

It lasted weeks, coupled with the exhaustion.

Then in March 1996, she got the rash—raised and bumpy, red, all over the upper parts of her limbs. Now she did go to the doctor, who told her: "I don't know what it is."

Linda pressed on. Still working sixty-five-hour weeks, with her husband seeing and raising her work hours, she had the newborn and her daughter and was now trying to be the type of mother she idealized. She'd have conference calls in her Ford Explorer with the kids in the back seat. In September 1996, she had the dinner party with colleagues from Boston Consulting Group when her left big toe exploded to golf-ball size.

Her doctors didn't know what it was. They speculated it was Lyme disease. They were wrong, but this is indicative of the mind-set of medicine: There must be a pathogen or foreign agent at work.

Two weeks later, her right big toe blew up the same way. Then it was her left knee—like a grapefruit.

Linda was under full-fledged assault. Her primary care doctors weren't sure why. No wonder. For as prevalent as autoimmunity is, its diagnosis can be worse than tricky. For a long time, the condition was invisible.

28

The Wolf

When a person goes to a doctor, he or she starts with the symptoms. *My throat hurts, my leg aches, I've got fever, there's this rash.*

The doctor starts with the symptoms too. The first question: *What's bothering you?*

With much of disease, the medical questions then move into the cause of the symptoms. *You have a cold, pneumonia, a virus or bacteria, a parasite, cancer.*

The thing about autoimmunity is that the questions and answers sometimes don't get any further than focusing on the symptoms. *My joints ache, I have fever, I've got this rash, I have diarrhea, constipation, mind-numbing fatigue.*

And the doctor says: *I believe you, but I can't find anything wrong.*

Something is wrong, all right. But there is nothing to point to. There is no pathogen. There is no infection. There is no foreign disease.

No aspect of the immune system story is as pointed or pure as that of autoimmunity.

The mystery began with the werewolf.

As early 963 AD, one history notes, scientists observed an unusual condition that left people looking as if they'd been bitten by an animal. Hippocrates was the first to describe symptoms consistent with the skin disease, and Hebernus of Tours is thought to be the first to apply the term *lupus* to it. The word derives from the Latin word meaning wolf. Its sufferers showed sores, "ill-favored lesions," and various other colorful descriptors—gnawing dermatosis—that I read in accounts of the medieval history of the illness. These "grotesque" lesions appeared on the face, lower limbs, all about. These symptoms—some caused by lupus and some not—were considered the product of a wolf bite and even a sign that someone had turned into a werewolf, according to the Lupus Endeavor, an advocacy project.

The vernacular and diagnosis were equaled in their primitive nature only by the treatment: "cutting away diseased tissue or burning it with caustic chemicals. These interventions rarely provided a cure, and patients suffered gradual disfigurement over decades," reads a case history of lupus that appeared in 2016 in the vaunted medical journal *The Lancet*.

In 1872, the Vienna School of Medicine employed a doctor named Moritz Kaposi, who associated lupus with other conditions in the body, including arthritis. In the latter half of the nineteenth century, a Canadian physician, Sir William Osler, connected lupus lesions with even more conditions, including impacts on the heart, lung, and liver. Dr. Osler gets credit for the name systemic lupus erythematosus.

The key word here is *systemic*. The condition was not just about the skin. Something bigger was going on.

On a parallel path, scientists had begun identifying and exploring an unusual condition that led to pain in the joints. In Paris in 1800, a doctoral student assessed nine patients and determined that the joint pain they were suffering was different from the overarching

A nineteenth-century woodcut depiction of a woman suffering from arthritis, long before the agony of women like Linda and Merredith was taken seriously by the medical community. (Wellcome Collection)

diagnosis of gout that many people suffered from. The student initially called it asthenic gout. Then in 1859, at University College Hospital in London, a pioneering doctor and researcher, Alfred Garrod, gave this condition its modern name: rheumatoid arthritis.

This was a disease characterized by inflammation, typically impacting the joints. Remember that inflammation is the body's own response to disease. Inflammation is not "other." It is "self."

Did that mean the disease was caused by self?

The very idea that the body would attack itself was still relatively new. The pioneering immunologist Paul Ehrlich introduced the term *horror autotoxicus* right around 1900. Autoimmunity. The body attacking itself.

As immunology spun forward into the twentieth century—a relatively tiny community in a field thought by many to be a backwater—the people exploring these unusual inflammatory conditions were an even smaller subset. One research hub was the Mayo Clinic in Rochester, Minnesota. In 1926, according to a Mayo history, 574 patients were admitted to the rheumatology service with joint swelling and pain. The presumption was that the cause was chronic infection—something foreign was sparking it. This was, of course, wrong. Vaccines were tried. They led to serious side effects, even death.

Imagine: an overheated immune system getting a boost from medicine and vaccine.

Other patients were treated with "fever" treatments—meaning that fever was induced to try to reverse the symptoms, an effort to stop a mysterious condition by literally igniting the immune system.

Then, in 1929, there came a revelation.

Doctors who work on joint pain are called rheumatologists. A pioneering rheumatologist named Dr. Philip Hench, working at the Mayo Clinic, noticed an oddity with one particular rheumatoid arthritis patient. Her joint pain and stiffness seemed to get better when she developed acute jaundice. She got a disease and her joint pain got better, not worse.

The doctor also noticed that other rheumatology patients saw their symptoms recede after surgery and during pregnancy. Dr. Hench theorized that the patients under duress had secreted a compound that countered whatever was attacking their joints, explains a history in the journal *Clinical Chemistry*.

Dr. Hench had a hunch. When patients experience stress and are under duress, it typically means that they are secreting adrenaline. Dr. Hench theorized that joint pain and inflammation were being dulled by a secretion from the adrenal gland, a small, triangular-shaped nub located atop each kidney that produces essential hormones. Fueled by the hypothesis, Dr. Hench and a biochemist named Edward Calvin Kendall made one of the most important discoveries in the history of autoimmunity.

In an effort to discover the substance that had improved the condition of these Mayo patients, Kendall started trying to isolate secretions from the adrenal glands of cows. The biochemist took regular shipments of adrenal tissues from Chicago slaughterhouses, according to the history published in the journal *Clinical Chemistry*. The biochemist discovered a handful of hormones that were labeled with letters of the alphabet: A, B, C, etc. The one that changed science was called Compound E.

It was studied initially because it seemed relatively simple. It also made patients feel better, sometimes euphoric.

It took many years to refine and isolate. Then in 1948 at the Mayo Clinic, the very scientists who had begun working there in 1929 gave Compound E to a twenty-nine-year-old woman immobilized with severe rheumatoid arthritis. "Two days and two more injections later, the patient could walk and left the hospital to enjoy a three-hour shopping spree," reads a recounting of the story published in the same 2010 scientific article.

"This startling result stunned people throughout the world," another history recounts. The two Mayo researchers won the Nobel Prize for this in 1950.

You might know Compound E by a different name, cortisol. Cortisol is a steroid that suppresses the immune system. Steroids are the first line of defense against many autoimmune disorders. They are a mixed blessing, as you'll see later on. For the moment, though, the discovery of steroids in the field of immunology and medicine was analogous to the discovery of a vaccine or antibiotic; they were a tremendous revelation, a response to a vexing problem, but a response that came without an understanding of the underlying mechanism of the disease they were meant to treat— autoimmunity.

As in so much of immunology, other major pieces began to fall into place in the late 1950s, as scientific technology improved. For instance, lupus researchers could now see that the condition involved a patient's own immune cells eating away at free-floating material in the bone marrow. This was a double whammy of sorts. The bone marrow helps gestate and stimulate the immune system, and it was under attack by the very system it had helped spawn.

Another major break in the autoimmune mystery came in the late 1950s and 1960s from Dr. Henry George Kunkel, widely considered one of the pioneers in the field of autoimmunity. Dr. Kunkel spent his entire career working at the Rockefeller Institute in New York. His patient and research base there included women suffering from liver disease. Many of the women also had arthritis. This was thought to be largely coincidental; after all, arthritis can have many causes, including aging and repetitive physical stress. It is not always an autoimmune issue.

In studying these liver patients, Dr. Kunkel isolated some of the women's antibodies—those large specialized molecules on cell surfaces that help our bodies target what to attack. Among the molecules he collected, Dr. Kunkel observed and isolated nineteen antibodies that did something quite disturbing. These antibodies, rather than picking up on and reacting to signals from foreign cells, reacted to the patient's own white blood cells.

Now he understood rheumatoid arthritis. He'd found a key test to prove the body was attacking itself, using the very properties that other immunologists had begun to understand as essential to defending ourselves against invaders. It was a brilliant and essential insight.

In 1948, a related test was developed to probe for the presence of antinuclear antibodies. These antibodies can bind to the nucleus of a normal cell and had been shown to be present in virtually everyone with systemic lupus. (Complicating matters, the antibodies also appear in people who don't have lupus, so initially, the test worked only about half the time; by the mid-1960s, the effectiveness of the test rose to 95 percent.)

Thus, at the dawn of the nuclear age, there were somewhat effective tests for only two of nearly a hundred known autoimmune disorders. And there was little in the way of treatment.

This was largely the state of affairs in the late 1960s when a patient in her forties came to Johns Hopkins suffering terrible joint pain, sobbing, trying to hold it together. Among others who tended the woman was a medical student named Bevra Hahn, who would go on to become a prominent specialist in this area.

The woman's story captures the reality of autoimmunity during the period. Despite all the fantastic science by Dr. Kunkel and others, autoimmunity remained difficult, if not impossible, to diagnose and treat. This challenge was compounded by the sexist way that women were viewed in society at that time. When women complained—whether about physical or emotional duress—they were often deemed "hysterical." Society could be quick to dismiss the work of women solely as caretakers of children and the home, employment deemed second class and not particularly taxing. In reality, this work could be brutal on the joints and compound the pain.

"Women had very defined roles. The husband never did the laundry. The husband never made a meal. Diapering a baby is really hard when your joints are swollen and painful," Dr. Hahn explained. This patient, a white woman from a middle-class family, wore pants, not skirts, to hide swollen joints.

Dr. Hahn didn't have much to give her. Steroids didn't work. "All I had was aspirin and gold shots," she explained. There was a theory, she told me, that compounds with gold in them could kill tuberculosis germs, and there was another theory that TB was related to autoimmunity. The treatment, as Dr. Hahn pointed out, "was very primitive."

In 1975, Carolyn Wiener, a behavioral scientist at the University of California at San Francisco, wrote a research paper that captures the reality of living with autoimmunity. The article is painful to read. It gives shape to the emotional side of living with a disease, rheumatoid arthritis, that is difficult to diagnose, with "no cure available."

The paper starts with a journal entry from a twenty-nine-year-old woman suffering from RA:

Being physically comfortable
And doing a simple chore

Can raise one's spirits to
Levels of supreme joy.

Persistent pain and wretched
Tiredness brings one to
Near despair

In the next forty years, I
Wonder how many variations thereof
I shall experience.

"Rheumatoid arthritis patients learn, along with their diagnosis, that the disease is not only incurable but that its specific manifestations are unpredictable. As often as not, they hear the physician say, 'You are going to have to learn to live with it,'" the paper reads.

Among the "self-doctoring" strategies that the paper describes for coping are "ingestion of celery juice, or massive doses of vitamin E or plastic bags filled with powered sulphur wrapped around the feet at night . . . a poultice of ginger root steeped in vodka and an alloy."

Another tactic in the paper is referred to as "covering up." Autoimmune sufferers would pretend they weren't suffering, try to look as if nothing were wrong. It was a mixed blessing. Friends and family would then assume nothing was wrong and expect full activity from the afflicted.

I was privileged to hear the intimate medical and personal narratives of two autoimmune sufferers, Linda and Merredith—two of the stories I share in this book. Their stories also provide insight into some of the key factors that impact the balance of everyone's immune system—namely, sleep, stress, hygiene, family history, and the ecosystem of our gut, known as the microbiome.

And they tell us about the fight of this ever-increasing group of patients to move out from the shadows.

29

Invisible Evidence

On October 10, 1996, Linda, her knee aching and grapefruit-sized, came to a rheumatologist's office in Palo Alto. She had an appointment with Dr. Rhonda Elaine Lambert, one of the region's best in the field. Dr. Lambert held an adjunct position on Stanford's faculty and served as a consultant to multiple sports teams, college and professional. She knew joints and her specialty was rheumatology.

She ran a battery of tests on Linda.

Linda's X-rays were normal. Her rheumatoid factor was negative. Her test for antinuclear antibodies, a sign of lupus, was negative.

"Her labs were unremarkable," Dr. Lambert said. Except for one number. Linda had taken a test to measure sedimentation rate, which provides a broad-based measure of inflammation. Her score should've been under 20. It was 94. Inflammation off the charts. Then there was the most obvious test of all, the eyeball test, the clinical exam. Linda had a grapefruit knee. Her joints ached. Her toes had exploded.

Dr. Lambert hesitated on a diagnosis. Even to this day, autoimmunity remains one of the most challenging conditions in medicine to diagnose with precision.

The Johns Hopkins University School of Medicine divides the materials of the diagnosing of autoimmunity into three categories that collectively sound like types of evidence at a criminal trial. The evidence can be direct, indirect, or circumstantial.

Direct evidence involves being able to transmit and reproduce the condition from one human to another—to, in effect, replicate the autoimmune process.

There aren't very many examples of this. The best involves a doctor in the 1950s who pursued a time-honored tradition in science: experimenting on himself. The doctor injected himself with the blood of a sufferer of idiopathic thrombocytopenic purpura, or ITP—a condition that causes excessive bruising and bleeding, leading to purplish spots or regions on everything from skin to tongue and lips. The condition is caused by a low level of platelets, which cause clotting, and the doctor and his colleagues surmised this was because the body's own immune system was attacking the platelets.

Within hours of injecting himself with the patient's blood, the doctor's platelet count plummeted and he had to be hospitalized. The result was so specific that it showed that an antibody in the woman's blood—an autoantibody—attacked a self-antigen. The condition was renamed thrombocytopenic purpura.

One reason such evidence is hard to come by is simply that you can't introduce a foreign body into a human, including another person's cells, without initiating an immune response. This is why organ transplantation is so challenging. Studying the mechanisms from human to human involves many complications.

So scientists pursued a second course, indirect evidence. This entails replicating a human condition in mice. This is doable with multiple sclerosis, where the immune system interferes with the central nervous system. It can be induced in mice by vaccinating the mice with an antigen that is much like the one that humans attack in themselves.

But direct and indirect evidence allow diagnosis of only a hand-ful of the autoimmune disorders. This leads to the heavy use of circumstantial evidence, which can be unsatisfying for patients and doctors. It involves looking at family history, the high lev-els of antibodies associated with the condition, and several other factors, including the circumstances that lead to onset, including stress.

There's another big factor: Is the sufferer a woman?

"Females make more of an immune system response than males do. We all know that," I was told by Dr. Hahn, the physician who treated rheumatology patients with gold shots in the late 1960s. Dr. Hahn rose to become the president of the American College of Rheumatology in the late 1990s, another glass-ceiling breaker in the field, and she is now chief of the division of rheumatology at the UCLA School of Medicine.

Women live longer, and they tend to be the last to die in, say, a famine or in an epidemic. The exact reasons aren't known, but Dr. Hahn offers some theories as to why, in an evolutionary sense, women might have a stronger immune system. One possibility is that women confer the first immunity to their babies. Indeed, as she says, "the baby's protection from disease is pretty much exclu-sively from the mother's immune system antibodies."

Another theory, she offers, "is that women tend to be care-givers." Women, by definition, are there when the baby is born, whereas the man might've flown the coop. A caregiver might need higher protection from disease. Women generally have more body fat than men, so perhaps they have more immune system cells, Dr. Hahn postulated to me.

She also noted that many of the genes that are associated with lupus and rheumatoid arthritis are on the X chromosome. (Women have two X chromosomes, whereas men have one X and one Y.) So the math of autoimmunity became greatly weighted toward females. (Another piece of science trivia: When researchers want to create an antibody to study, they use a female animal, not a male. You get more antibodies.)

A woman's relatively elevated immune system "is associated with living longer. But you have higher antibodies. That might make you sick and cause you to die," Dr. Hahn said. What an incredible trade-off: longer life thanks to powerful defenses that can turn on themselves! This is an extraordinary insight too into the larger balance struck by our elegant defense. When the system contributes to longer life, it comes with a powerful potential cost. More defense, more risk. In day-to-day terms, the downside of a strong immune system, then, is that it can become more susceptible to being inflamed or set off by lack of sleep or stress or—this will likely go without saying—genetics. Fifty percent or more of cases appear to have a distinct genetic link, with a family member having had the condition, or a related one.

Another factor that can throw the immune system out of whack is infection. Say, for instance, a pathogen invades the body. The immune system then responds and succeeds in eliminating the pathogen. But this response can spur autoimmunity when the immune system doesn't fully shut down and remains in hyperdrive, even though the pathogen has been ousted from the Festival of Life.

These are the same kinds of mechanics, incidentally, that make smoking such a risk for rheumatoid arthritis. Smoking introduces all sorts of foreign particles into the body, sucked down the throat and into the lungs, turning the immune system into a busybody surveying the particles and damage. In the case of rheumatoid arthritis, a possible cause is smoking—"an enormous possible trigger," explained Dr. Lambert, Linda's doctor.

Linda's case didn't offer Dr. Lambert much in the way of direct or indirect evidence. The circumstantial evidence spoke volumes. She wasn't a smoker, but she had a host of other risk factors.

Inflammation. Check.

Infection. Check. Prior to onset of arthritis, she'd had strep—a disease that might've set her immune system off and running.

Sleeplessness, Check.

Stress. And then some.

Linda had first seen Dr. Lambert on October 10. She returned two weeks later. This time, Dr. Lambert took virtually one look at her and just knew.

Linda had to be moved into the clinic in a wheelchair. Multiple joints were now inflamed. "Her disease had taken off like a rocket," Dr. Lambert said.

At this point Dr. Lambert was sure Linda was suffering from rheumatoid arthritis. The doctor prescribed a first-line treatment of steroids. Specifically, she gave Linda a drug called prednisone. Dr. Lambert described it as "being like a big hammer. It shuts a lot of things down."

It's used to treat many inflammatory diseases. "But unfortunately it has all this spin-off all over the body." Such as weakening your immune system, leaving you susceptible to infection, and making it even harder to sleep. That's partly because it interacts with the adrenal gland.

"We really don't like to use prednisone in the long term."

Dr. Lambert felt she had no choice in Linda's case because the damage to Linda's joints had progressed so quickly and was so extreme that it could have become irreversible. "She likely would have ended up in a wheelchair permanently."

The steroids put Linda out of balance. She couldn't sleep at night, so she took Ambien and then a drug called Flexeril, a muscle relaxant, to stay asleep. That was the bad news.

The worse news is that the steroid regimen wasn't working—not well enough.

Her hands hurt so much, she couldn't button her pants. She started wearing pull-on pants. One day, when she dropped her daughter off at school, another little girl came up to her and in all innocence asked, "How come you always wear the same clothes?"

Linda couldn't use her hands to pick up her infant son and would try to grasp him with her forearms. She wore gloves when she went out in case she had to shake hands with someone, to soften the impact.

When Linda came back in December 1996, Dr. Lambert drained 65 cubic centimeters of fluid from her left knee (about 65 teaspoons) and 30 cubic centimeters from her right knee. She was on 30 milligrams of prednisone, and the tablets are available only in 20 milligrams.

By now Linda was also taking a drug called methotrexate, which was originally used in chemotherapy for blood cancers, aimed at interfering with malignant white blood cells. But white blood cells are immune system cells, so when they are attacked, the body becomes highly vulnerable to infection.

"I had an eye infection, an ear infection, a yeast infection, a bronchial infection—in every orifice or opening you could get an infection, I had one. I was a petri dish. I was kind of thinking that the swelling was better than this."

By the spring of 1997, Linda was on fifteen medications—some to help with the autoimmunity, some to stem the activity of the other medications.

Then, as things were seemingly close to being under control, another trauma hit.

Linda had gotten enormous help from her mother-in-law throughout the prior six months. That April, her mother-in-law committed suicide. A lifeline was gone, and Linda's marriage began to deteriorate. It's not an exaggeration to say that as Linda's immune system lost balance, her life fell out of balance too.

As the drugs began to take effect, lessening the rheumatic symptoms, her immune system continued to wrestle with basic challenges. In late summer of 1997, a major client wanted her to come to London. The anti-inflammatory medications had weakened her immune system to the point that she was suffering from a terrible cough. In London, she went one night to see a play called *Art*. She took a pillow with her to the theater to cough into.

One day she met with her client's European president. She was supposed to be advising him, but all she could do was cough. She excused herself. In the hallway, she tried to gain control. But for twenty minutes she coughed. "I couldn't go back in the room."

Linda was making excruciating trade-offs with her immune system, suppressing it at great cost. But medicine was on the verge of snagging this problem by the tail.

30

Best of Both Worlds (Sort Of)

In November 1998, the U.S. Food and Drug Administration approved one of most anticipated drugs in medical history: Enbrel. It was aimed at treating rheumatoid arthritis.

What was so widely anticipated about Enbrel, made by a Seattle company called the Immunex Corporation, was that it was designed specifically to limit the effects of an overreactive immune system without undermining the entirety of the system.

It was built around the kind of discoveries made with monoclonal antibodies in the seventies. The ability to isolate and replicate individual proteins was allowing drugmakers to develop medicines constructed around very specific molecules. These proteins, sometimes antibodies, injected into the body, would theoretically attach to and react with only very specific cells in the body.

For instance, Enbrel works by using proteins to interact with a particular cytokine—an immune system signaler—known as tumor necrosis factor, or TNF. What TNF does is send a signal that causes a cell to die, specifically by experiencing apoptosis. This is a crucial normal process in our Festival of Life, and it is quite elegant and orderly. A cell receives a signal to die, essentially to kill itself, and it begins to break into little digestible chunks that then get eaten by the janitors, the macrophages. (*Apoptosis* comes from a Greek word meaning a falling off.)

With Enbrel and other drugs that act on TNF, the idea is to get the cells that are causing problems to commit suicide. Obviously, getting malignant cells to off themselves can be useful in cancer. In the case of rheumatoid arthritis, it's also advantageous to have overzealous immune cells commit apoptosis. Instead of attacking Linda's body, the cells would kill themselves.

(Heady stuff, and even weirder given another bit of trivia: The protein used in Enbrel is produced in hamster ovaries.)

Dr. Lambert couldn't wait for Enbrel to come out. "It was a game changer. We all knew that. We were waiting for it."

In early 1999, Linda took her first infusion of Enbrel via a shot in her upper thigh. It took several months to work, and then . . . whoa.

The swelling began to subside. The pain began to diminish.

Enbrel wasn't scorching the earth of Linda's immune system, as steroids could do, but acting in a more targeted fashion. This was part of the dream of immunology, going back to Jacques Miller—to understand the immune system well enough to tinker with it.

"My immune system is allowed to work, and this binds to the parts of my immune system that are attacking me and neutralizes them," Linda said, sounding awed. "Once I went on this drug, my life just changed."

Enbrel is now one of the bestselling drugs in the entire world. It generated $5.5 billion in sales in the 2017 fiscal year for Amgen, the company marketing it.

The story of how these drugs work is even more sensational when it comes to cancer, and I'll tell you that story shortly.

But this is not pure miracle. Autoimmunity is too complex for one size fits all, and the new drugs still leave many people feeling invisible.

This brings us to Merredith Branscombe—both an echo of Linda and a study in contrasts.

31

Merredith

Merredith, born just two years after Linda and nine hundred miles away in Denver, woke up one morning in 1977 with a low-grade fever. Her joints hurt. It felt like a vise was crushing them. Merredith began to experience mysterious symptoms intensely through her teens—a low-level surfacing of aches, pains, and fevers largely left untreated. The doctors thought she might have mono.

Merredith was in much the same boat as her mother, who would experience unusual spells, body pain, and swelling, and often had trouble digesting her food. Merredith remembers her mother with her hand held over her forehead, feeling faint. Was she ill? It was hard to pin down. Maybe it was her childhood, and the secret, and the stress of a life spent feeling like an outsider.

Merredith's family lived in a neighborhood called Park Hill. In the late 1960s, the area was white but integrating. Neighbors didn't like that. On multiple occasions, Merredith's family came home to find flyers posted on their house urging them to leave before the colored folks arrived.

Merredith's parents disagreed with this bigotry. Her father, an editor at *The Denver Post*, did some research showing that property values actually went up, not down, when neighborhoods integrated. There was more demand for homes. Merredith's father wrote the first editorial in the paper urging integration. The next

Merredith Branscombe. (Courtesy of Merredith Branscombe)

night, a Molotov cocktail was thrown through their window. It was as if the white people in the neighborhood were overreacting to the presence of something they perceived as other. What was alien? What was self? The country wrestled with itself and integrated.

It was all very personal to Merredith's mother, Bea. She had been raised Catholic, then Congregational; she had married an Episcopalian and worked with black and white churches to foster integration. She also worked in the civil rights division of the Colorado state government. She'd learned the hard way to fight for integration.

Merredith's mother had made a narrow escape from the Nazis. At its core, the experience was a story of how the body politic can overheat and turn self into other.

In Austria, Bea's grandfather had been a baron and the personal physician to the kaiser—in effect, Austria's surgeon general. His son, Paul Von Domeny, played an important societal role too as a doctor and World War I hero.

They were also Jews. When anti-Semitism took root after World War I, the family converted to Catholicism to avoid persecution. In the end, though, assimilation didn't work.

The Nuremberg Laws were passed in 1935, denying basic political and social rights to Jews, who were identified not by religious belief but by bloodline. This nationalism proved to function as a kind of autoimmune disorder: Hitler was attacking productive, healthy, essential parts of the whole of Germany and Austria. On Kristallnacht, in November 1938, Merredith's mother saw her father and mother taken into the street, put on their hands and knees, and made to lick shards of glass from broken windows.

Beatrice, Merredith's mother, told her this one night when she was ten and they had watched a documentary on Kristallnacht. Merredith's mother never drank alcohol, and she never cried. This night, she did both.

Merredith's mother told her: "My parents were told they were Jewish vermin and had to clean up the street. I have never forgotten it. My mother, my *beautiful mother*, they made her lick until blood was gushing from her mouth. I was terrified."

Merredith asked, "Are we Jewish?"

"We were Jewish enough."

As the war started, Merredith's mother and her parents narrowly escaped to London, where her father's business had an office.

They changed their name again to Sutton, so it would be neither German nor Jewish. With her parents, Merredith's mother lived in London during the Blitz, volunteering as a Girl Guide—part of the group that was a precursor to the Girl Scouts—leading other girls into the tunnels to survive the bombing. Her grandfather, Paul Von Domeny, died in Theresienstadt, a concentration camp, in March of 1944, a victim of Hitler's autoimmune machine.

Perhaps it is no wonder Merredith's mother got sick.

When Merredith was a little girl in Denver, her mother had joint pain, exhaustion that felt like brain fog, and gastrointestinal troubles. Merredith and her sisters made fun of her because they didn't know any better.

"I think about my mom sometimes and I cringe, with that little guilt," Merredith reflected. Her mom took supplements and all kinds of pills, whatever the doctors told her to take to feel better. "She was in pain, and nobody had an answer."

It wasn't until the early 1990s that it was confirmed that her mother was suffering from ulcerative colitis and Guillain-Barré syndrome, a rare and nasty disorder. Here, our elegant defense, the immune system, turns against the lining that coats the ends of long nerve cells that extend along the periphery of the body. The linings of the nerves, known as myelin sheaths, are crucial because they help the body to quickly and efficiently transmit information by insulating these cells and, in effect, keeping out other information. In Merredith's mother, though, T cells and B cells had begun to cooperate in attacking the myelin sheaths.

"Guillain-Barré is called a syndrome rather than a disease because it is not clear that a specific disease-causing agent is involved. A syndrome is a medical condition characterized by a collection of symptoms," notes a description of the illness furnished by the National Institute of Neurological Disorders and Stroke. More evidence of the fact that there's no bad guy to identify. Just self, turned inward.

Further proof of the autoimmune mystery comes from the same organization: "No one yet knows why Guillain-Barré—which is not contagious—strikes some people and not others. Nor does anyone know exactly what sets the disease in motion."

The sum of this history left Merredith genetically predisposed to autoimmune disease. Then she began to experience her own trauma.

In her junior year of college at Northwestern, where she'd gone on scholarship, she was raped. She was devastated, and the assault, like so many such college incidents, went unaddressed. She returned home and never went back. And she was still dealing emotionally with a previous sexual assault. When she was fifteen, she was assaulted by a priest in the family's church. She'd had a cold at the time, and he said he'd make her some soup. Instead, he climbed on top of her, pinned her down, tried to kiss her, stuck his tongue down her throat. She escaped, but was left wondering if there was something about her that made her a victim, made her needs invisible.

Merredith, in an email, wrote me:

Since I have told you about having a priest try to seduce me, and the collective shrug from the powers who putatively were supposed to protect me, and the same at Northwestern, I hope you understand when I say it was the same feeling: the people I trusted were fallible, maybe doing the best they could/not intentionally causing harm, but I was not important enough to cause them to vary from business as usual.

During this period, her physical symptoms became worse and finally exploded.

In the summer of 2001, Merredith and her family went to Playa del Carmen, a resort town south of Cancún, Mexico. One day they swam in the Cenotes, a series of exotic, expansive underground caves. When they got home, Merredith felt feverish and achy. Her joints hurt, the pain was excruciating—but she assumed it was just from the fever. "My head was so swollen that the top of my skull was spongy."

Her fever spiked at 103.

She went to a doctor, who ran some tests. There was no infection. A fever of 103, and no pathogen!

A different doctor called her back. "I'm really sorry," she recalled that he told her, "but you have lupus." The test had shown that she had a particular antibody that was indicative of lupus—the antinuclear antibody—at more than ten times the normal levels. But it wasn't yet proof.

She was so naive about lupus. "I thought, well, at least it's not disfiguring." She laughed, looking back.

Merredith was referred to a clinic in Denver and to a specialist named Dr. Kathryn Hobbs. After a few visits, Dr. Hobbs changed Merredith's official diagnosis to rheumatoid arthritis, chiefly because there were more drugs approved for a diagnosis of RA than for one of lupus.

Merredith's course of treatment closely followed the one given to Linda.

Merredith began taking steroids, the first of many drugs that would do as much damage to her as they did healing, leading to fatigue, infection, fever. The steroids made her feel worse. In 2002, she took methotrexate, the cancer drug intended to interfere with production of certain cells by depriving them of vitamin B. One of its "benefits" is that it suppresses the immune system. Its side effects outweighed the benefits for Merredith, who used it for two months.

It was the same with another drug she tried early on, azathioprine, which suppresses the immune system by interfering with cells at the DNA level and has a host of side effects, including some long-term elevated risk of cancer, according to the American College of Rheumatology.

In 2003, Merredith began taking Enbrel, the wonder drug.

It worked great for a while. Until it didn't. Her symptoms got worse.

There were, by now, other options. A competitor of Enbrel was called Remicade, made by Janssen Biotech and approved by the FDA in 1999. It also worked by blocking TNF. But it wasn't cheap; a story published in the *New York Times* when the drug was approved reported that a single treatment of Remicade cost $9,500, less than the $11,400 single-cost treatment of Enbrel.

In Merredith's case, the only one profiting from her treatment was the medical industry. Monoclonal antibodies didn't work for her. To ease the agony, she juggled heavy pain meds, like Vioxx and Celebrex, along with tramadol. More drugs, more immune system imbalance, no relief. Bloody stools, rashes, bouts of debilitating pain, and fevers, and not a damn pathogen to blame.

Then Merredith's own rheumatologist, Dr. Hobbs, came down with strange autoimmune system symptoms. It seemed to be arthritis of the spine. The doctor had to begin her own treatment. It didn't work. Dr. Hobbs started getting ulcers on the outside of her body—attacks that were much more pronounced than a rash, as if the skin was being eaten away.

She had developed a very rare, dangerous autoimmune condition known as pyoderma gangrenosum. It appears to involve massive stores of tumor necrosis factor going on the attack against self.

Dr. Hobbs started getting care from a dermatologist named Dr. Meg Lemon, who had vast experience in internal medicine. Dr. Lemon, coincidentally, had also been consulting on Merredith's case.

Dr. Lemon strongly suspected that Merredith had dermatomyositis. It is a relatively rare condition, characterized by rashes and muscle weakness. But Merredith's biopsy was negative for the condition, and she didn't have the blood markers that accompany it.

In the office, "I had trouble convincing other people," Dr. Lemon told me of her diagnosis of Merredith. But Dr. Lemon could see the evidence with her own eyes. She could see the rashes, and she knew of Merredith's regular experiences with pain and weakness. "I saw her rash, and I said: This is what you've got."

In the end, though, Dr. Lemon conceded that Merredith is a classic example of something she sees all the time with autoimmunity. "We listen to their stories and we try to fit them into the box, but millions of people don't fit into a box. They're not making up what they're experiencing, they're not flakes. We just don't know what's wrong yet."

For these people, Dr. Lemon said, "science has not caught up yet."

At some point, hopefully soon, the cause of the symptoms will become clearer, and a more specific treatment than Humira or Remicade or steroids will arise. Dr. Lemon noted that the progress in the last few decades has been immense. Hope is real, and reason for optimism great.

But it is also time to see the invisible women. Their plight is genuine. Witness what happened to Dr. Hobbs, Merredith's rheumatologist.

"It was one of the most hideous fucking cases," Dr. Lemon said.

Dr. Hobbs's neutrophils began to eat away at her skin. Skin is the first layer of the immune system; it's the shield. She tried all kinds of treatments to stop the attacks.

Merredith, who by now had become friends with Dr. Hobbs, said Hobbs's treatments became a mixed blessing. They offered hope of slowing the immune system, but the double-edged sword of reducing her defenses meant that when infection came, Dr. Hobbs was less able to fight it off. Dr. Hobbs texted Merredith pictures of boils on her body.

In February 2015, she texted Merredith: "This is pretty much the most awful thing that's ever happened to me. I can't stop crying. I'm so scared about all these docs thinking that I'm going to die." In March, Dr. Hobbs texted Merredith that she was close to getting sepsis, which is a systemic infection that gets into the bloodstream and overwhelms the body.

On October 9, she texted Merredith: "I'm sorry, M. So very sick right now. Trying to stay out of hospital. I have more bad bugs. I take IV antibiotics four times a day."

Widely beloved, Dr. Kathryn Hobbs died on October 25, 2016, done in by her own immune system and the impossible challenges of threading the needle of trying to treat it.

"She died," Dr. Lemon said, "of absolutely the most hideous autoimmune disease."

By this time, Merredith had spent more than a decade looking for a way to put the brakes on her immune system. It ravaged her physically, emotionally, spiritually. She lathered on medicines intended to slow the attacks of her immune cells on her joints and GI tract, her skin and her heart muscle. The medicines left her open to regular infection. She was a walking pharmacy, an alphabet soup of drugs. This is the detailed list she made for me of drugs that she was taking in 2014 or had regularly taken prior to that:

- Steroids (I could take them only when there was an actual infection; not sure why)
- Methotrexate
- Imuran
- Enbrel—injections, for about a year, maybe two
- Medications for pain and other side effects:
  - Opioids (stopped taking them for pain a few years in)
  - Bextra, Vioxx, Celebrex
  - Adderall (as needed, for brain fog)
  - Tramadol (as needed, for pain)
  - Topamax, Neurontin—these are anti-seizure medications; really not sure why they prescribed them.
  - Valium, cyclobenzaprine—to help with sleep. Exhaustion often forces patients like me to self-medicate with caffeine, but then we can't sleep. I ended up finally quitting the sleep meds because I was too spacy afterward, and if I combine that with lupus brain fog, it's not a good scenario for a patient who is still trying to function and work. But I probably took cyclobenzaprine, in particular, for a decade, on and off.

She couldn't tell what made her worse, her condition or the meds. Her own rheumatologist had an autoimmune disease and

died from side effects of the same medication that had been prescribed for Merredith.

In late 2015, Merredith had awakened with a new set of symptoms. She's a terrific writer, and nothing I can write can compare
with the poignant email she sent me about what happened next:

> I was sitting on my bed in the late morning after the third
> consecutive night of pain so bad it awakened me from a drugged
> sleep. I had returned from Mexico and *again* was sick, but these
> were new symptoms. I was exhausted, exasperated, desperate,
> and hoping I'd find something that would add some context or
> insight to what was going on.
>
> The house was quiet; I was trying to work from home since
> I was too tired, and in too much pain, to go into my office. I had
> been up most of the night trying whatever I could: stretching, pain
> pills, massage, a hot bath, a cold bath—but the pain remained, like
> someone had plunged knives into both sides of my body and was
> just . . . turning and driving those knives deeper and deeper into my
> muscles. There was no relief, no matter what I tried.
>
> I needed a solution: I still have a business to run and kids to
> parent, so I couldn't drug myself senseless. That morning I turned
> to Google as kind of a Hail Mary: these symptoms were both new
> and excruciatingly painful, and I wanted to preemptively check on
> whether the new pain was maybe a part of my condition or a side
> effect of treatment. My rationale was, if I was going to call my
> doctor, I should check first that this wasn't "part of the condition," as
> I had heard so many times across the years. I typed in "minocycline
> + autoimmune," thinking maybe I'd see side effects or protocols.
> I expected to find something reassuring; and instead I found
> something called "minocycline-induced autoimmune syndrome."
> In short, what I was taking could either cause my condition or
> make it worse. I remember thinking, "What the actual f**k?" as I

skimmed through those abstracts, one after the other. My doctors had prescribed chronic minocycline and told me it was "less toxic," but either hadn't bothered to look up the studies, didn't know about them, or didn't care—statistically either it worked or it didn't. But what if the other things I was doing—avoiding sun, avoiding sugar, etc.—were what was helping, and minocycline was either *not* helping or making it worse?

What came to my brain, unexpectedly unburied, was this line from Yeats's poem, "Among School Children":

How can we know the dancer from the dance?

In my journey since my diagnosis up until that day, I had been a dutiful patient. I had done what they asked, minus the steroids and methotrexate family because I had unhelpfully (to them) pointed out that they made me feel worse. In just my mother, myself and my daughter, that's three generations of women who trusted Doctors/Medical Advances to help them even as the help made them worse.

I felt like I was stepping off a path—almost physically. Alone. **I was not going to be protected or saved.** It sounds melodramatic, but it didn't feel urgent or exciting. Instead, it was just that other options had been eliminated. I could either keep taking something that had demonstrably also *caused* my disease, move on up the healthcare ladder to even more horrific treatments, like Rituxan, while waiting for the next Breakthrough Drug to wend its way through the FDA . . . or I could try to help myself.

I remember looking skyward that day, one of those Boulder early winter days with coolly transparent sunlight, to where my mother must be, or must symbolically be. I asked her, out loud: "Is this tough enough yet?"

I felt immeasurably sad, but not terrified. Looking back, I think it's because having one path closed off to you is weirdly freeing. It's just . . . math.

A new journey began for Merredith, an experiment that was day one. Was there some way she might find to save herself (not that she hadn't been trying already)? Merredith went back to basics, with diet, lifestyle, and a bunch of other natural methods—she researched meticulously—that offer some clues to the way many of us can find balance. (For instance, she takes vitamin D, because she can't have sun, and a cocktail of supplements—C, B, iron, CoQ10—that have proven less toxic to her.)

Dr. Lemon has much to say on this topic. Some of it sounds counterintuitive, but now that you know more about the immune system, and its delicate balance, it should make lots of sense.

Dr. Lemon said it is not uncommon for patients with strange rashes or other unusual symptoms to come into her office with a widely used refrain: "They say their immune system is weak. They've gone down the rabbit hole of the Internet, reading people who proclaim themselves experts telling them to boost their immune systems. When people tell you your immune system is weak, they are wrong. Anyone who wants to boost your immune system doesn't know what they are saying."

Or rather, not in the way they mean it.

Dr. Lemon thinks one great way to keep your immune system in balance is to . . . eat the food you drop on the floor. Her philosophy, as she puts it, is that people need to stop oversanitizing their world so that their immune systems are introduced to lots of bacteria, parasites, and other pathogens and can react to them as millions of years of evolution have refined them to do.

This philosophy is increasingly widely held. It is called the hygiene hypothesis, and the broad idea is that we are starving our immune systems of training and activity by an excessive obsessive focus on cleanliness.

"I tell people, when they drop food on the floor, please pick it up and eat it. Get rid of the antibacterial soap. Immunize! If a new vaccine comes out, run and get it. I immunized the living hell out of my children. And it's okay if they eat dirt. We have animals in our homes, and they sleep with us. If your dog shits on the floor, clean it up, of course, but don't use bleach. You should not only pick your nose, you should eat it."

Seriously?

Yeah, Dr. Lemon says, why not?

"Our immune system needs a job. We evolved over millions of years to have our immune systems under constant assault. Now they don't have anything to do."

Our elegant defense has grown restless.

"But it's a hard discussion to have with patients. They've been brainwashed that they have a weak immune system. People look at me like I'm crazy."

The numbers of people suffering some common autoimmunity and allergies have increased sharply.

Evidence is mounting about how the balance of our immune system has changed—how the modern world has upset it.

Is Dr. Lemon crazy? Should you pick your nose?

Here I will briefly turn the focus to four major factors of day-to-day life that impact autoimmunity and immunity broadly, in the lives of Linda and Merredith, Bob, Jason, and you and me. These four factors are sleep, stress, the gut, and hygiene.

All these roads will eventually lead us back to Jason, and the epic battle waged in his life's festival.

32

Should You Pick Your Nose?

Don't laugh. These are now serious questions. Should you pick your nose? Should your children pick their noses?

"I don't know. It might have some negative social consequences," one epidemiologist told me. She was quite serious: The biggest downside to nose-picking (and eating) might be the social consequences. Could it actually be a health advantage?

Should your children eat dirt? Maybe.

Should you use antibacterial soap or hand sanitizers? No.

Are we taking too many antibiotics? Yes.

For more complete answers, let us turn to nineteenth-century London.

The British Journal of Homeopathy, volume 29, published in 1872, includes a startlingly prescient observation about hay fever: "Hay fever is said to be an aristocratic disease, and there can be no doubt that, if it is not almost wholly confined to the upper classes of society, it is rarely, if ever, met with but among the educated."

Hay fever is a catchall term for seasonal allergies to things like pollen and other airborne irritants. This nineteenth-century essay, incidentally, says it can be difficult to distinguish between hay fever and asthma or rheumatism. This is worth noting because these turn out to be autoimmune disorders, and allergies wind up as a close cousin. The immune system is overreacting.

With this idea that hay fever was an aristocratic disease, the British scientists were on to something.

More than a century later, in November 1989, another highly influential paper was published on the subject of hay fever. The paper was short, less than two pages, in *BMJ*, titled "Hay Fever, Hygiene, and Household Size." The author looked at the prevalence of hay fever among 17,414 children born in March of 1958. Of sixteen variables the scientist explored, he described a "most striking" association between the likelihood that a child would get hay fever allergy and the number of his or her siblings. It was an inverse relationship, meaning the more siblings the child had, the less likely it was that he or she would get the allergy. Not just that, but the children least likely to get allergies—also known as atopic diseases—were ones who had older siblings.

The paper hypothesized that "allergic diseases were prevented by infection in early childhood transmitted by unhygienic contact with older siblings or acquired prenatally from a mother infected by contact with her older children.

"Over the past century declining family size, improvements in household amenities, and higher standards of personal cleanliness have reduced the opportunity for cross infection in young families," the paper reads. "This may have resulted in more widespread clinical expression of atopic disease, emerging in wealthier people, as seems to have occurred in hay fever."

This is the birth of the *hygiene hypothesis*. It provides one of the most telling and vivid insights into the challenges that human beings face in our relationship with the modern world. In a nutshell, that challenge revolves around the idea that we evolved over millions of years to survive in the environment around us. For most of human existence, that environment was characterized by extreme challenges, like scarcity of food or food that could carry disease, unsanitary conditions and unclean water, withering weather, and on and on. It was a very dangerous environment, a heck of a thing to survive.

At the center of our defenses was our immune system. These defenses are the product of the millennia of evolution, the way a river stone is shaped by the water rushing over it and the tumbles it experiences on its wayward journey downstream.

Along the way, we humans learned to take steps to bolster our defenses. Prior to the discovery of medicines, we developed all manner of custom and habit to support our survival. In this way, think of the brain—the organ that helps us develop habits and customs—as another facet of the immune system. For instance, we used our collective brains to figure out effective behaviors. We started washing our hands or took care to avoid certain foods that could be dangerous or deadly. Some cultures avoid pork, which is highly susceptible to trichinosis; others banned meats, with their toxic loads of E. coli. Ritual washing gets mentioned in Exodus, one of the earliest books in the Bible: "So they shall wash their hands and their feet, so they will not die."

Our ideas evolved, but for the most part, our immune system did not. This is not to say that our immune system didn't undergo change. The immune system responds to our environment and learns. This is central to the branch of the immune system known as the adaptive immune system. Our immune system comes into contact with various threats, develops an immune response, and then is much more able to deal with that threat in the future. In that way, we adapt to our environment.

But adaptation is not the same thing as evolution. Adaptation involves responding to the environment within the limits of individual physical capacities. To take a random example: If you learn that you are more likely to catch a bird if you hunt at dawn, you will wake up early and go hunting. You are adapting to your environment. By contrast, evolution involves fundamentally changing our physical capacities over the course of many generations. Evolution, in this case, might optimize our bird-catching ability by the development of wings. For humans to become winged creatures, it would take eons.

What does this have to do with your immune system and allergies? Lots.

To survive, we *adapted* within our physical capacities. We washed our hands, swept our floors, cooked our food, or avoided certain foods altogether. We learned and adapted.

Then our learning and adaptation began to intensify as we built quickly upon past discoveries. Human discoveries came in leaps and bounds. We developed medicines like vaccines and antibiotics. Virtually overnight, we changed the environment with which our immune system interacted. We improved the hygiene of the animals we raised and slaughtered for food, and that of our crops and kitchens. Particularly in the wealthier areas of the world, we purified our water, developed plumbing and water and human waste treatment plants; we isolated and killed bacteria and other germs. But for the most part, our immune system continues to be the same one humans have always had. It had developed and evolved to allow us to survive in a particular type of environment—one teeming with pathogens. On one level, we had given a major helping hand to our immune system. Its enemies list was attenuated. On another level, though, our immune system is proving that it cannot keep up with this change.

At a core level, we have created a mismatch between our immune system—one of the longest surviving and most refined balancing acts in the world—and our environment. Thanks to all

A 1921 Lysol Disinfectant ad. Germ killing has been great for business, mixed for public health.

the powerful learning we've done as a species, our immune system isn't getting the regular interaction with germs that helped to teach and hone it—that "trained" it. It doesn't encounter as many bugs when we are babies. This is not just because our homes are cleaner, but also because our families are smaller (fewer older kids to bring home the germs), our foods and water cleaner, our milk sterilized, and on and on.

What does an immune system do when it's not properly trained?

It overreacts. It becomes aggrieved by things like dust mites or pollen. It develops what we called allergies, chronic immune system attacks—inflammation—in a way that is counterproductive, irritating, even dangerous. There has been a rise in autoimmunity too.

The numbers are significant.

The percentage of children in the United States with a food allergy rose 50 percent between 1997–1999 and 2009–2011, according to the Centers for Disease Control and Prevention.

Of similar magnitude, the jump in skin allergies was 69 percent during that period, resulting in 12.5 percent of American children with eczema and other irritations.

In keeping with the themes mentioned earlier in this chapter, food and respiratory allergies rose with income level. More money, which typically correlates with higher education, meant more risk of allergy. This could reflect differences in who reports such allergies but also differences in environment.

These trends are seen internationally too. Skin allergies "doubled or tripled in industrialized countries during the past three decades, affecting 15–30 percent of children and 2–10 percent of adults," according to a paper published by the British Society for Immunology. Asthma, the paper reads, "is becoming an 'epidemic' phenomenon."

By 2011, one in four children in Europe had an allergy, and

the figure was on the rise, according to a report by the World Allergy Organization. Reinforcing the hygiene hypothesis, the paper noted that migration studies have shown that some types of both allergy and autoimmunity rise as people move from poorer to richer countries. The prevalence of diabetes is higher in Pakistanis who move to the United Kingdom than in those who remain in Pakistan. The incidence of lupus is higher among African Americans than West Africans, the paper notes.

There are similar trends with inflammatory bowel disease, lupus, rheumatic conditions, and in particular, celiac disease. That entails the immune's system overreacting to the protein molecule in gluten. This attack, in turn, damages the walls of the small intestine. This might sound like a food allergy, but it is different in part because of the symptoms. In the case of an autoimmune disorder like this one, the inflammation happens in the area of insult; the immune system attacks the protein and associated regions.

Allergies can generate a more generalized response. A peanut allergy, for instance, can lead to inflammation in the windpipe, known as anaphylaxis, which can cause strangulation.

In the case of both allergy and autoimmunity, though, the immune system reacts more strongly than it otherwise might, or than is "healthy" for the host (yeah, I'm talking about you).

This is not to say that all of these increases are due to better hygiene, a drop in childhood infection, and its association with wealth and education. There have been many changes to our environment, including new pollutants. There are absolutely genetic factors as well. But the hygiene hypothesis—and when it comes to allergy, the inverse relationship between industrialized processes and health—prevails.

An instructive study has to do with the Amish.

The Amish are not known for a tendency to lend excitement to most proceedings, but this is the kind of study that gets researchers all fired up. The study looked at the prevalence of allergy among two communities, one Amish in Indiana and the other Hutterite in South Dakota. Why is this particular study so exciting to scientists? It's because these two groups have remained relatively isolated since they moved to the United States several hundred years ago (the Amish in the 1700s from Switzerland, and the Hutterites in the 1800s from South Tyrol, which borders Switzerland in northern Italy). The upshot: They descend from relatively similar genetic stock and have like-minded approaches to things known to impact allergy, including large family size, high rates of vaccination, and, notes the study, "taboos against indoor pets." Ah, but livestock. No taboos there.

This is both a similarity and the key difference.

The Amish in the study "practice traditional farming, live on single-family dairy farms, and use horses for fieldwork and transportation. The Hutterites live on large, highly industrialized, communal farms."

There is another major difference, this having to do with the prevalence of allergy. Only 5 percent of Amish schoolchildren suffered asthma, compared to 21 percent for the Hutterites.

On a slightly lesser measure of sensitivity—called allergic sensitization—7 percent of Amish kids qualified, compared to 33 percent for Hutterities.

The researchers asked what was causing two groups of people with very similar genetic backgrounds, similarly isolated from other groups culturally and environmentally, to have such different allergy profiles.

One powerful clue the researchers discovered was that the households of the Amish were much more likely to have allergens, "from cats, dogs, house-dust mites, and cockroaches." Forty percent of Amish homes had them, compared to 10 percent for the Hutterites. You'd think you'd rather live in the Hutterite household, right?

We're just getting started.

Residue from bacteria, the kind that cause disease, was nearly seven times higher in the Amish homes.

Now the twist. Researchers looked inside the bodies of the Amish subjects and found evidence that turns knee-jerk disgust on its head. The Amish children had a higher proportion of immune system cells called neutrophils. Remember these? They are front-line fighters.

Among the Amish, there was also a relatively lower proportion of eosinophils. These are another kind of white blood cell; they are solid, all-purpose fighters essential to destroying viruses, bacteria, and parasites. They can cause inflammation, and that, as you know now, is a double-edged sword. In fact, they are highly associated with allergy and autoimmunity when these numbers are elevated; in excess, they can be markers of asthma and eczema, lupus, Crohn's disease, and other conditions.

The Amish and Hutterites were subjected to a type of bacteria known to elicit a strong immune system response measured by the levels of cytokines, like interferon, interleukins, and such. Overall, the bacteria elicited the same twenty-three cytokines, but in lower proportions in the Amish.

"As compared with the Hutterites, the Amish, who practice traditional farming and are exposed to an environment rich in microbes, showed exceedingly low rates of asthma and distinct immune profiles that suggest profound effects on innate immunity," reads the paper in *The New England Journal of Medicine*.

The researchers then did studies in mice to try to reproduce the results. They did studies that showed that mice raised in relatively microbe-rich environments, like the Amish, developed immune systems that were more effective in key ways than mice raised in Hutterite-like environments.

I'm going to quote the study in all its scientific-vernacular glory, partly because readers who have come this far have earned the ability to grasp most of it.

> The concordance between findings from studies in humans and in mice was remarkable: in both studies protection was accompanied by lower levels of eosinophils, higher levels of neutrophils, generally suppressed cytokine responses, and no increase in levels of T regulatory cells or interleukin-10. Thus, the finding that these features were largely dependent on innate immune pathways in mice suggests that innate immune signaling may also be the primary target of protection in the Amish children, in whom downstream adaptive immune responses may also be modulated.

Now for the plain-English version. The dust and pet filth, the cockroach miasma, and the barnyard residue, far from being an enemy, impacted the immune system through both pathways, innate and adaptive, and the Amish kids were far less likely to get an allergy.

So maybe you should pick your nose and eat what you extract? The study doesn't speak to that. But it might explain the urge we sometimes get. Maybe we're shoving a few germs up the nostril to test the system, the same way that kiddos put lots of things in their mouths. During the research for this book, a well-known immunologist told me that children should "eat a pound of dirt a day." He was being somewhat glib, but you can now get his point.

A lot of products have been marketed to suggest otherwise.

When I was a kid, I collected Wacky Packages. They were packs of trading cards and stickers that made fun of major brand names. Milk-Bone for dogs was rendered as Milk Foam, and Band-Aid was Band-Ache. Each pack came with a rectangular stick of pink gum that was almost certainly manufactured in the 1700s.

Among the products that made the cut for Wacky Packages were multiple hygiene and cleaning products: Windhex (Windex); Ajerx (Ajax); Toad (Tide detergent).

No wonder. These kinds of products were heavily advertised during a surge in hygiene-related marketing that began in the late 1800s, according to another novel study published in 2001 by the Association for Professionals in Infection Control and Epidemiology. You heard right; researchers from Columbia University, who did the research, were trying to understand how we became so enamored with soap products. Some highlights:

- The Sears catalogue in the early 1900s heavily advertised "ammonia, Borax, and laundry and toilet soap."
- "During the early to mid-1900s, soap manufacturing in the United States increased by 44 percent," coinciding with "major improvements in water supply, refuse disposal and sewage systems."
- The marketing trailed off in the 1960s and 1970s as antibiotics and vaccines were understood to be the answer, with less emphasis on "personal responsibility."

- But then, starting in the late 1980s, the market for such hygiene products—home and personal—surged 81 percent. The authors cite "return of public concern for protection against infectious disease," and it's hard not to think of AIDS as part of that attention. If you're in marketing, never waste a crisis, and the messages had an impact. The study cites a Gallup poll from 1998 that found that 66 percent of adults worried about virus and bacteria, and 40 percent "believed these microorganisms were becoming more widespread." Gallup also reported that 33 percent of adults "expressed the need for antibacterial cleansers to protect the home environment," and 26 percent believed they were needed to protect the body and skin.

They were wrong.

It's not just the public who have been mistaken in our perceptions. Many doctors have been misled or are being plain irresponsible when it comes to a related topic: the use of antibiotics.

I've already described antibiotics as a marvelous, world-changing advance. At the same time, the vast unnecessary prescription of antibiotics is bad for the individuals who take drugs they don't need—it kills important bacteria in their bodies—and worse for society as a whole. What is happening is that bacteria are evolving, with rapid-fire speed, such that they can survive antibiotics. The bacteria that survive are called superbugs. They sound apocalyptic, but they are very, very real.

A report published in late 2014 found that 700,000 people die annually from common bacterial infections that have grown resistant to drugs.

Of course the bacteria evolve to be resistant! Like any creatures, bacteria mutate, and the mutated bugs that are resistant to drugs are the ones most likely to survive. This is as basic as science gets.

And bacteria are coming into contact with antibiotics all over the place. Not only are they among the most prescribed drugs in the world, but they are used widely around the world to fatten chickens, pigs, and other livestock. The use of antibiotics for meat allows much faster delivery and growth, creating cheaper protein. That is a big deal, particularly in developing countries. But the use of antibiotics is hardly limited to countries with emerging economies; the United States in 2015 sold 34 million pounds of antibiotics for use in "food-producing animals," according to the FDA. That was around 80 percent of the antibiotics used in the United States altogether.

Heavy use of antibiotics worldwide is putting enormous pressure on bacteria to evolve. And scientists have discovered that the bacteria are evading antibiotics more quickly than previously expected because of the way they are evolving. Bacteria are passing back and forth among themselves a genetic code that allows them to fend off attacks from antibiotics. In fact, bacteria that are under attack from antibiotics can effectively call out to their fellow bacteria for help ("Send me some protective genetic material!") and the resistance can be transferred.

The report that came out in 2014 predicted that by 2050, 10 million people will die from resistant bacteria annually, which by that point will be more than the 8.2 million people predicted to die of cancer that year. There is a case to be made that this is among the top three medical crises facing our world, as widespread and shared as climate change but with much more immediate impact.

A scientist who headed up efforts at the World Health Organization to develop world policy to limit use of antibiotics told me that philosophically, there's a lesson that goes counter to a century of marketing: We're not safer when we try to eliminate every risk from our environment.

"We have to get away from the idea of annihilating these things in our local environment. It just plays upon a certain fear."

How easy is it to play on our fear of having more bacteria swarming around the insides of our bodies?

In fact, more bacteria may be exactly what we need.

Here we return briefly to Linda Segre, the golfer who got terrible rheumatoid arthritis. Just two years ago, as her life was well back on track—she had become an executive vice president of Diamond Foods—she received an unusual request from an elite group she belongs to made up of high-level executives. The group's aim is to have the executives communicate, share their wisdom and experience, and also to keep them abreast of cutting-edge issues in the world. That includes their health.

The group sent out a message explaining they were going to ask members if they'd like to check on the health of their gut. Of course, thought Linda. So she did what they asked, sending them a fecal sample.

This act goes to the heart of another key aspect of the increase in allergy and autoimmunity. It also concerns our overall health and the balance of the immune system.

Meet your friendly neighborhood bacteria: the microbiota.

33

Microbiome

At least half of the cells in our body are bacterial, not human. One hundred trillion bacterial cells, and they are mostly in our gut. In the individual, they are called a microbiota, and the collection of them, and the breadth of their genetic building blocks, is called the microbiome.

One review on the subject written by scholars at the University of Colorado at Boulder reported that there are 3.3 million microbial genes in the human gut, "compared to roughly 22,000 genes present in the entire human genome." Another study estimated there were 1,000 species of bacteria in the gut with 5 million genes. The scope of the microbiome is, in a word, huge.

The paper from Boulder notes that human beings have virtually the same set of genetic material—you and I are 99.9 percent similar in our underlying genetic building blocks. But the microbiome—the underlying genetic material of the bacteria in our gut or hand—can differ by 80 to 90 percent. (Worth noting is that most bacteria are in your gut, though there are also 500 bacterial species in your mouth, and about the same number in your "airways"—the respiratory system; 300 million are on the skin; for women, about 150 million are in the genital infrastructure.)

"Everything you're looking at is covered in microbes. You just can't see them. They colonize the world but are invisible to us," explained Sarkis Mazmanian, a Caltech professor who is seen as one of the field's leading thinkers. Back when Mazmanian first got started—and "developed this love for bacteria"—he thought they were "these insidious little creatures that want to make us sick. I was wrong."

For the longest time, there was a theory that the reason we could coexist with the bacteria in our gut was that there was a protective layer lining our gut that acted as a powerful barrier. The barrier, a mucus-like lining with the texture of Vaseline, behaves as something like a force field between the small and large intestines and the rest of the body. This lining, it was thought, kept our microbiota from getting into the rest of the body and thus away from the immune system. This theory is called immunological ignorance.

The immune cells, in effect, were thought to be ignorant of these bacteria among us.

This thinking was incomplete, if not outright incorrect. Mazmanian and others have since found that the gel that lines the gut is colonized by the microbiota, and their presence puts them very much in close proximity to cells that can trigger an immune response. On the other side of that gel-like wall is a line of cells, called epithelial cells, that is heavy with immune triggers.

This suggests the microbiota have developed with the deliberate ability to interact with and stimulate our immune system.

To make sense of this, step back and think about humans in the context of the world. We exist in a literal and metaphorical sea of bacteria. We must coexist with them in the same way that we must coexist with each other. Imagine if you were at war all the time with your neighbors; you'd eventually kill each other off, as surely as the Hatfields and the McCoys. Instead we find common ground, cooperate, and maybe get some help with coexistence by setting

A salmonella bacterium (upper right) detected by epithelial cells in the digestive tract. (David Goulding/Wellcome Trust Sanger Institute)

up fences and boundaries. The relationship between humans and bacteria is even more intimate than that. We are different from one another, we can harbor antagonistic feelings at times, but for the most part we are highly supportive of each other, which is essential for protecting one another's survival.

Still, from an evolutionary perspective and in the scale of epochs, the initial meeting of immune system and bacteria was not a friendly one.

"The first encounter was likely antagonistic—until a truce was reached," Mazmanian says. The immune system and the bacteria felt each other out and through evolution made a mutually beneficial peace. Then they determined that they could survive only together; they could serve one another's aim to survive. Mazmanian calls it a "partnership—both players on the same side of the net, fighting common foes."

The shared foes are the handful of pathogens—the bacteria and viruses and parasites that would kill the human tissue. In the grand scheme, these pathogens are a tiny fraction of the bacteria in the world. For the bacteria we cooperate with, our microbiota, these pathogens become a common enemy because our body is the host where the microbiota lives. "The bacterium collaborates with the immune system to fight off an invading microorganism," Mazmanian said. "It makes sense as a benefit to both."

That evolutionary perspective plays out on an individual stage. Each of us develops a working relationship with our environment. It's a social contract of sorts with the bacteria in our midst, and the contract is highly personalized and highly variable. This is underscored by one powerful piece of scientific trivia: The microbiota in the gut of an infant who is delivered vaginally differs from the microbiota of an infant delivered by cesarean section. In the early days, our business partner, the microbe, then goes from 0 to 60. It's put nicely in an article coauthored by, among others, a Stanford geneticist:

> The microbial colonization of the infant GI tract is a remarkable episode in the human lifecycle. Every time a human baby is born, a rich and dynamic ecosystem develops from a sterile environment. Within days, the microbial immigrants establish a thriving community whose population soon outnumbers that of the baby's own cells. The evolutionarily ancient symbiosis between the human GI tract and its resident microbiota undoubtedly involves diverse reciprocal interactions between the microbiota and

the host, with important consequences for human health and physiology. These interactions can have beneficial nutritional, immunological, and developmental effects, or pathogenic effects for the host.

That dense but highly informative passage refers to bacterial colonization of the baby's digestive tract. The colonists are "microbial immigrants"—another indication of the essential balance and blurring of self. Who are we and what is other? What is alien? To survive, how essential is it that we cooperate with other and not shun or destroy?

The paper makes several other powerful scientific points. One of them is about the role that our environment plays in the formation of our microbiota. In mice, babies who live in the same cage with their mothers have microbiota more similar to their mothers than do babies of the same mother who are caged elsewhere. As the paper notes: "The bacterial population that develops in the initial stages is to a significant extent determined by the specific bacteria to which a baby happens to be exposed."

To explain why these bacteria are so crucial, I will briefly reprise a pioneer from the 1970s, Susumu Tonegawa. Tonegawa helped discover that the underlying genetics of the human immune system can be so diverse because genes rearrange at random during development and infection so that each of us comes equipped with an immune system with a great ability to recognize and then "bind" to a wide range of potential threats. We have developed a nearly infinite array of antibodies; to many, this was the key to how we could survive.

However, despite the profound and extensive nature of the tool kit, it is not sufficient to ensure our survival. Here is where the microbiome comes in. "The human genome is not sufficient to confer all the benefits of health. We require input from the microbiome. We need this second genome. So we actually harbor two genomes, our own and our microbiome," Mazmanian told me.

The incredible cooperation between human and microbe has led to a new terminology to describe us. We are superorganisms. Yes, that's the scientific term. You should feel good about yourself. You are superpowered, a human being strengthened by the power of bacteria.

But what specifically does the microbiome help with?

Digestion, nutrition, obesity—broadly, how much energy we take from foods and how effectively we squeeze nutrients from them—but also anxiety and mood, and significantly in this context, how we defend ourselves against pathogens and defend against ourselves.

It helps to see the idea in practice.

One of the many variations of T cells, we now know, is called a T regulatory cell, or Treg. It is a powerful subset of our T cells that has been shown, among its other roles, to help suppress the immune system. In the big picture, this makes sense; it is part of a defense network that is primed to destroy party crashers without being so overzealous that they ruin the party.

In that respect, Treg cells are not so unusual. What makes them worthy of note here is that there is a decent chance they wouldn't exist without the presence of the microbiome in the gut. What Mazmanian discovered, using experiments in mice, is that Treg cells don't get developed when certain gut bacteria are missing. In other words, when the microbiome of the mouse is incomplete, so is the immune system.

Mazmanian and his fellow researchers also discovered that the bacteria had a signaling mechanism that stimulated the development of the Treg cells. The way this works, simplistically, is that the gut bacteria send a message that is passed via the immune cells that line the gut and then is received by cells in the bone marrow or thymus that are awaiting the command to take on the Treg identity.

Mazmanian described the upshot to me in fairly stark terms. "There are entire cell types in the body that don't exist because the DNA doesn't have all the information to tell that cell to develop," he said. It's not just Treg cells, but natural killer cells and other killer immune cells that appear to be triggered by the bacteria.

Broadly, Mazmanian's work also shows that the microbiome plays a key role in *dampening* the immune system, in addition to helping it to attack foreign invaders. This is because—as I hope has become abundantly clear—the immune system is as dangerous to us as it can be to invaders. The microbiome can't afford to have the host hurt by its own immune system, by an overzealous police state. It is in the microbiome's self-interest to keep the body from attacking itself, so the bacteria contribute to helping keep the immune system in check.

"The immune system is a loaded gun, and when it fires and is uncontrolled, then you get allergies, then you get autoimmunity, then you get inflammation," Mazmanian says.

There is a powerful punch line to Mazmanian's work: The way we relate to bacteria in the world dictates our health. If the relationship gets out of whack, our immune system becomes unbalanced too. "What we are discussing here," Mazmanian says, "is the modern interpretation of the hygiene hypothesis."

The hygiene hypothesis stated that our environment has become so clean that it has left our immune system insufficiently trained. Mazmanian and others believe that the microbiome lies at the heart of the challenges faced by the immune system in modern times.

Our efforts to scrub our environment of bacteria, while well intended, have wound up limiting the number of bacteria that colonize and populate our gut. Mazmanian jokes that the toilet is a mixed blessing, compared to pooping in the woods: half burying the bacteria, half washing our hands. Instead of that, "We're flushing the good guys away."

Does he mean it when he says we'd be better off with fewer modern amenities?

Well, it is true that people in less developed countries, say some parts of Africa, have much more complex microbiomes than we do. When Mazmanian first got to Caltech in 2006, an idealistic part of him thought that these complex microbiomes and the environment that fostered them are superior to those of the Western world.

"Trust me," a colleague told him, "you don't want a microbiome full of tropical virus and parasites."

There are also lots of examples where exposure to dangerous pathogens at a young age can lead to illness later, or to autoimmunity. *So it's not that we want to do away with many modern amenities and be surrounded by bacteria.* But it is true, Mazmanian says he has since learned, that the result of overly cleansing our environment and using antimicrobial soaps and wipes is to limit the microbiota we pass back and forth among us. This is the essential point. As a species, we have all kinds of beneficial bacteria. Some of us are colonized with certain types of bacteria while others of us carry different ones. Throughout human history, we have passed these back and forth, shared them, creating a vast trading network through handshakes, hugs, and cheek pats, shared use of stair banisters or countertops, and on and on. Now we kill our microbiome instead of sharing it.

"We've distanced ourselves from infectious agents, but also distanced ourselves from microbes that confer benefits," Mazmanian said. "I will likely have a less complex microbiome than my mother, and that of my kids will be less complex than mine. Every generation may just be less diverse."

We relied on these microbes to fill out our defense, including the signals that dampen our immune system. This appears to be one key reason allergies and autoimmunity have risen. We aren't getting the signals that say: Slow down. Don't react to the pollen. Don't attack yourself.

These ideas, the hygiene hypothesis and the microbiome, impact the health of the population at large, the broader environment surrounding our immune systems.

We have reached an inflection point at which our relationship with bacteria is fundamentally shifting. Bacteria are organisms we share this planet with, and with which we have coexisted for millennia. The relationship is changing because we as a species are fighting to survive and bacteria are fighting to survive. That relationship has always been in flux, but the pressures on the relationship have intensified because of human technology, like antimicrobial soaps, antibiotics, and nonorganic foodstuffs. These advances, wonderful in certain respects, hallmarks of human innovation, are so powerful that they have sharply thrown off the tenuous balance between bacteria and us. This is similar in key respects to other technological advances that have had unintended consequences. The birth of the automobile allowed fast transportation and led immediately to thousands of road deaths; processed food allowed widespread preservation and transport of calories but led to junk food and a deadly obesity epidemic; cell phones changed communications overnight and have also threatened focus and attention, led to distracted driving and compulsive computer use; and so forth.

There is one key difference between those situations and those we confront with bacteria. In all those other cases, we have ultimate control. We can change how we behave or modify the technology. But in the case of our relationship to bacteria, we control only half the equation. We can try to take steps to put less pressure on bacteria, but we can't ultimately dictate how these powerful organisms will react.

So what does this add up to? First, simply, we must have an awareness that we are sharing the earth with these distant cousins. Second, we will need societal policies to address things like the resistance of bacteria to antibiotics. We can at least try to use our technology more judiciously.

On an individual level, there is less we can do. But it is possible to be less neurotic about the use of products that can have a counterproductive impact on our own bodies and on bacteria as a whole. We can choose to eat foods not raised with antibiotics. We can decide to pick up a piece of food dropped from the floor, rinse it, and eat it.

I concede that it's a tough balance to strike at times. After all, the rise of resistant bacteria makes it hard to think about eating food off a hospital floor (where bacteria are rampant) or eating meat without caution while traveling in a developing country when those meats might well have been raised without antibiotics and could be undercooked, allowing them to transfer resistant bacteria.

In the end, another major collective step we can take is to support science. It has yielded terrific answers and likely holds the key to helping us right the balance with our bacteria.

Meanwhile, on an individual level, there are other areas of life where we can take concrete steps to keep our immune systems in balance. These are central to health, not just autoimmunity, and they are under our control. In fact, if you pick only two chapters in this book to read and apply to your life, these are the ones. They focus on stress and sleep and the science of how they impact your immune system.

34

Stress

The stress we put our bodies under impacts the delicately achieved underlying aim of our immune system—to achieve balance. When I think of how delicate this balance is, I sometimes picture an elite gymnast on a balance beam leaping into a flip and landing, again and again, with no margin for error. Stress can act like a midair push, adding a jostle to an already precarious feat.

Credit for revelations of the role of stress is owed to a colorful couple at Ohio State University, Janice Kiecolt-Glaser and her husband, Ronald Glaser, who did seminal research by posing a question that you may well have asked yourself: How come you get sick after final exams?

The Glasers met on October 3 in 1978 at a faculty picnic at Ohio State University. She was junior faculty in psychology and he was then the chair of the Department of Medical Microbiology and Immunology. She was twenty-seven, and he thirty-nine. He'd already been twice married and divorced. On their first date, they went to lunch, and then he took her to his office, presumably to show off what a big shot he was. She noticed his unusual taste in art: A picture of a sperm hung on the wall, and on his desk there was a stand with a dried piranha.

Ronald Glaser and Janice Kiecolt-Glaser. (Courtesy of Kiecolt-Glaser)

"How can you trust a man with a sperm on his wall, who has been married twice?" She laughed, recounting the story.

As they got to know each other, he proposed that they combine their expertise—hers in psychology and his in immunology. "He thought it would be interesting," Kiecolt-Glaser recalled. "I didn't even know what a lymphocyte was."

To this point, little had been done on the subject of stress and the immune system. One early study had found that Swedish volunteers subjected to seventy-seven hours of noise and sleep deprivation suffered ill health effects.

And there was an "odd study," Kiecolt-Glaser recalled, performed at West Point. It had taken place in the mid-1970s and involved looking at which cadets were more likely to contract infectious mononucleosis, one of the eight types of human herpes viruses. They can be relatively harmless and are among the most common viruses in the world, but this class of virus obviously has a more troublesome side too.

The research took place over four years and involved a class of roughly 1,400 cadets. When a cadet entered West Point, he was tested to see whether he produced antibodies to fight Epstein-Barr virus, the pathogen that causes infectious mononucleosis. In other words, the researchers were testing whether the cadets' bodies had been exposed to the Epstein-Barr virus and had developed a defense that recognized it.

When the cadets entered, roughly 30 percent lacked the antibody for Epstein-Barr virus. They had essentially not encountered it in any meaningful way. Of that group that had the antibody, 20 percent eventually "became infected," according to the research paper, written by Yale scholars and published in 1979 in *Psychosomatic Medicine.* Among the cadets who became infected, 25 percent not only had antibodies but showed clinical signs of being sick. What was surprising was one common thread among cadets likely to develop infectious mono: They were doing poorly in school, had highly accomplished fathers, and were themselves particularly motivated to succeed.

They were "guys who were not doing well and really wanted to," Jan Kiecolt-Glaser said, "Ambitious guys, struggling in school, with a successful dad." Stress seemed to be playing a key role in the immune system's response.

"Ron said, 'Let's try a study with medical students.'" Herpes was the perfect test.

Not only is herpes among the world's most common virus families—nearly all adult Americans have been infected with several of the eight by age forty—it also has a very delicate, even profound, relationship with the immune system. It's a relationship that tells us something about how our defenses have evolved to make a calculation about when to attack and when to stand back. Sometimes when a pathogen is detected but the pathogen appears not to be spreading and not to be too dangerous, our elegant defenses watch and observe—acting more like peacekeepers than assassins. Herpes is a wonderful example.

It's worth noting of herpes that the genetics of the virus share key characteristics with human DNA. Notably, both have double-stranded DNA—the famous double helix. That's one way that herpes can be a bit flummoxing to the immune system, which is always scanning for self and alien. In this case, it can be hard for the immune system to identify the nonself.

Also, once a person is infected, the virus does something else that makes it challenging for the immune system. Herpes essentially lies dormant. For instance, the oral version tends to hang out in cells inside the roots of nerves at the base of the skull (or in other nerve roots near and around the spine). Frankly, I find this to be terrifying—a virus hanging dormant like a pod in the movie *Alien*.

Meanwhile, the immune system, sensing that something is afoot but not active or not so clearly different from us, shows up and hangs out itself, keeping watch. The presence of immune system cells essentially keeps the herpes in check. "The cops show up and keep the party down," said William Khoury-Hanold, a Yale immunologist. The herpes is effectively held in check.

Sometimes, though, when the immune system gets preoccupied, stressed, or tamped down, it provides an opening for the virus to emerge. Herpes, sensing this temporary weakness, travels down the nerve roots into the mouth and attacks. Now the Festival of Life is under attack and the immune system must respond in force.

This scenario makes it a terrific test case for stress and the immune system because it allows researchers to see what happens when a person experiences stress: Do our defenses get distracted such that the herpes virus can emerge from its ganglion hideout?

The seminal 1982 study done by the Glasers involved seventy-five medical students. The study measured the level of the subjects' natural killer cells and their antibodies. The students were tested prior to exams, then on final-exam days, and then, sometime later, when they returned from vacation.

"When Ron saw the results, he didn't believe it." After the exam, "the antibody levels were so high he didn't trust the data."

The numbers were even higher for the lonelier students. "The stress of exams was bad for everybody, but worse for the lonelier students," Kiecolt-Glaser explained.

There also was a severe reaction involving the natural killer cells too. Remember that these are among the first-line defenders in the immune system—the heavy artillery. During exams, there was a sharp *fall* in the number of natural killer cells that circulated outside the bone marrow.

The stress of exams was suppressing a key part of the immune system. Why might this be?

During exams, there is a surge of adrenaline. This precedes the release of steroids.

You already know that steroids dampen the immune system and are used to fight autoimmunity. There is a profound logic behind this relationship—between adrenaline and steroids and the immune system. It is crucial to our survival.

Steroids play a number of pivotal roles in allowing us to survive moments of acute stress. Crucially, for instance, steroids help to maintain the integrity of blood vessels; in times of stress, when the blood vessels might constrict, these steroids keep them intact and, without putting too fine a point on it, maintain your blood circulation and pressure so you don't faint and die.

These kinds of steroids also have a wavelike impact across our immune system as they circulate through the body. Virtually every cell has a receptor for these types of steroids, and that receptor is called a glucocorticoid receptor. When the steroids become active or elevated, they can reach many, many cells—"every cell in the body," according to Dr. Jonathan Ashwell, an expert in cell biology at the NIH. It's a remarkable idea in and of itself. In the giant festival, this hormone courses through the entire confines of the party and has an impact on the behavior of many, many partygoers. At least the ones that are self.

Cell receptors for steroids are outside their nucleus (the innermost part). But when the steroid reaches them, a reaction begins that moves the steroid into the nucleus. There it begins a process by which it interacts with the cell's DNA to change the proteins made by the cell. Chief among the influences made by these steroids is that "you repress expression of a lot of genes important for an immune response to occur," Dr. Ashwell explained.

Now why would it benefit us to have our elegant defenses repressed?

The logic comes again from evolution. If your ancestor was under sudden and profound stress—say, fearing a bear or lion attack—it would be problematic for inflammation to create fatigue or fever. For most of human history, stress meant imminent threat, and imminent threat meant the body needed to be alert, fully functional, even a bit superpowered. This is where the hormone cortisol comes in. It is secreted by the adrenal gland.

In times of stress, the release of cortisol comes slightly after the release of one of two other key hormones, norepinephrine and epinephrine. Steroids and these other two hormones are separate but highly related pathways in the stress experience. The first— the release of epinephrine and norepinephrine—is known as the sympathetic response, and it involves the central nervous system. The second—release of cortisol—takes a bit longer to cascade; it goes from the brain to the pituitary and adrenal glands and releases glucocorticoid, a natural immune system dampener.

In times of acute stress, if there's a virus in your system, the virus fight can wait. The bigger threat has teeth and runs a 3.5-second forty-yard dash.

Immune system responses have "a substantial energetic cost and the potential for collateral damage," according to Dr. Michael Irwin, the Cousins Professor of Psychiatry and Biobehavioral Sciences at UCLA's David Geffen School of Medicine and director of the Cousins Center for Psychoneuroimmunology. Dr. Irwin is also one of the foremost experts in the world in the connection between your immune system and your brain and behavior, including stress and sleep. The collateral damage is fever, fatigue, actual swelling, or inflammation—all things that might encourage someone to slow down and rest. Not good when you're up against the lion.

This relationship between adrenaline and the immune system has another key driver: It is highly regulated by how we sleep.

35

Sleep

You can sleep when you're dead. So goes an old adage that should be scratched from your vocabulary.

Sleep accounts for a quarter to a third of your life, and for good reason. Much about sleep remains mysterious, though current theory suggests that among the benefits is the fact that the body uses sleep to flush toxins from your brain. This is, in its own way, a broader immune system function of cleaning the Festival of Life of detritus. There are myriad other health benefits of sleep. Improved memory, cognition, and mood; less inflammation, and you know now how huge that is. Or, if you prefer to look at the other side of the coin, people who get insufficient sleep can put their health at tremendous risk.

Sleep problems predict death.

People who experience prolonged sleep disturbance are more likely to die, and to die earlier, than people who don't. "The effect sizes are comparable to other known risk factors, like being sedentary, being overweight, having depression," said Dr. Irwin.

Studies in lab animals have shown even clearer links between sleep and health. Sleep-deprived rats die.

In humans, sleep problems are rampant, as recounted in a recent paper by Dr. Irwin. Roughly 25 percent of Americans have sleep problems, and "insomnia is one of the most ubiquitous complaints in psychiatric populations," the research shows, at least telling you that you're not alone.

The risk of early death from sleeplessness was shown with considerable confidence in a 2010 research roundup of sixteen studies involving 1.3 million subjects. Among those studies was one that found the optimal sleep level for longevity was seven hours, with a particularly heightened risk of death for those who slept fewer than 4.5 hours. (This behavior is rampant. A poll published in 2008 found that 44 percent of adults were getting less sleep than they said they needed, around seven hours, while 16 percent got fewer than six hours.)

Curiously, the same 2010 study also found an increased risk for people who reported sleeping more than 8.5 hours. I asked Dr. Irwin about that figure, and he said it still wasn't well understood. "It's really been debated a long time."

The theory has been that when people slept longer, it was indicative of an underlying medical condition that eventually led to premature death. But Dr. Irwin said a close look at the studies doesn't bear that out. While Dr. Irwin is doing ongoing research to get at the answer, he does have a hypothesis. He thinks that people who report sleeping longer are actually people who are spending more time in bed but not truly sleeping longer. He thinks they fundamentally have a "sleep maintenance" problem that is more akin to getting insufficient sleep, and so they spend many hours in bed to overcompensate.

The bigger issue is sleeplessness.

The work of Dr. Irwin and others has shown that when it comes to the dangers of sleeplessness, all roads lead through the immune system. "The effect of sleep on the immune system is the critical link that's driving that risk."

I mentioned the sympathetic response, the flight-or-fight response, above. It has a powerful impact on heart rate, blood pressure, the flow of digestive juices, and other core involuntary functions. When we sleep, the system slows markedly and the norepinephrine and epinephrine turn off. "When we don't sleep, that activation continues at daytime levels," Dr. Irwin said.

His research has also shown that people who are deprived of sleep experience a decrease in activity of natural killer cells "to the same level of people who are depressed or stressed." So sleep sets off and intensifies an adrenaline-fueled dampening of the immune system.

Other research shows that sleep loss leads to specific changes to at least ten interleukins, along with other inflammatory processes, and studies show that response to vaccines is diminished in people with sleep deprivation, suggesting our immune systems don't learn as well when we are tired. People who don't sleep are more likely to develop heart disease, cancer, and depression. "We now have compelling evidence that, in addition to cognitive impairment, sleep loss is associated with a wide range of detrimental consequences, with tremendous public-health ramifications," one recent paper found. I like the direct language used in a different scientific paper that looked at what happens to rats when they are sleep deprived. "There was failure to eradicate invading bacteria and toxins."

It might come as little surprise that a healthy immune system helps promote or mediate sleep, with various studies showing that several key cytokines—those immune-system signals—can promote sleep. This happens when you're healthy but also when you are sick, or getting sick; then, your immune system sends stronger signals creating a feeling of fatigue, telling your body to rest, and creating more resources to fight infection. All of this means the relationship between sleep and the immune system is a tight and circular one.

Also, simply, lack of sleep often is caused by stress and leads to more stress.

So you get stressed, don't sleep, your sympathetic response kicks in, your immune system gets dampened, and the cycle, through more stress and less sleep, spirals. Alone this is key. But Dr. Irwin offers a fascinating nuance.

He believes that only *part* of the immune system gets dampened by this cycle. Stress and lack of sleep, Dr. Irwin believes, make it tougher to fight viruses but easier, or at least less difficult, to fight bacteria.

His theory makes perfect sense from historical and evolutionary standpoints. Picture your forebear facing an acute threat—say, an attack from a lion or bear or from a fellow human with a spear, or simply having injuries from a fall or scratches from a rock or bush. The immediate threat would come from a puncture wound or bite and the bacteria that might be transferred by that injury. So it stands to reason that the immune system would favor lending its limited resources to a bacterial response over a viral one.

To be clear, the release of cortisol can dampen both types of immune responses—to allow us to remain alert during an acute threat—but Dr. Irwin sees the dampening as having a greater impact on a response to viruses.

In either case, virus or bacteria, these primitive responses can wind up having a perverse effect in the modern world. After all, these primitive systems kick in all the time, as if the body were responding to an attack by a lion or bear, but the actual threats are much different today, and often much less dangerous.

"Those same systems of alarm and threat can become activated in a social situation. You get into an interpersonal situation, an argument with your boss at work," said Dr. Irwin. "The sympathetic nervous system gets hijacked. It's just as if we were exposed to an acute threat in Neanderthal times and were hurt."

Often, Dr. Irwin said, the culture adds another layer, pushing us forward, rather than letting the system settle down through withdrawal or sleep. "It's a badge of honor to see how little sleep you can get by on. If you can sleep less and maintain function at work, you're a better professional. You're a better human being. That crazy logic has led to a sleep-deprived society, and that is having huge health consequences."

As concerns autoimmunity, there hasn't been a large study that has expressly tested the relationship among stress, sleep, and a hyperactive immune system, but Dr. Irwin says "there's a good case to be made" for a link between sleeplessness and autoimmunity. At the very least, the indirect connection speaks for itself: Lack of sleep leads to stress and vice versa, creating a vicious cycle that disregulates the immune system.

Dr. Lemon, the Denver physician who treated Merredith and believes strongly in the hygiene hypothesis, says she tells patients worried about their immune systems, "Your job isn't to keep your house spotlessly clean. You should sleep until you're not tired anymore. Sleep is the easiest medicine to regulate. A single night alters your immune system. It blows things out of whack in one night."

She says that she is by no means blaming people who get diseases like autoimmunity or cancer for their stress or sleeplessness.

Sometimes disease just happens.

It happened to Jason. He's our last story, the most telling of all about this extraordinary moment in time when we're putting to work nearly a century of understanding the balance of our immune system.

Part V

JASON

36

A Word About Cancer

In late summer of 2010 Jason was diagnosed with Hodgkin's lymphoma.

It's a cancer of the immune system. The word *lymphoma* refers to the lymphatic system, the network of nodes where the immune cells gather. In Hodgkin's—named for the nineteenth-century English doctor who discovered it—B cells have mutated into malignancy.

Cell mutations take place all the time inside the body. All of us get cancers. You might well have one now. Most of these mutations die off, simply because they are too mutant to survive or because the immune system identifies them as alien and destroys them. In the case of Hodgkin's, the cancer takes advantage of the immune system, dupes it, and even uses it to thrive.

The cancer cells "look like self in disguise," said Dr. Alexander Lesokhin, an expert in blood cancers and a hematologic oncologist at New York's Memorial Sloan Kettering, one of the leading research institutes in the world. Part of the way that Hodgkin's and other cancers disguise themselves is by tricking the T cells that would ordinarily help kill off the mutation. What the cancer does is send a signal to the T cell to self-destruct.

Why would the T cell do that? Why would it even have such a receptor on its surface capable of receiving a self-destruct signal?

It's because the immune system has many mechanisms that are

aimed at slowing it down, shutting it off, keeping it from overheating. Cancers take advantage of these fail-safe mechanisms to survive.

The self-destruct receptor on the T cell is called programmed death. For short, PD.

On the cancer is a molecule called PDL-1, a programmed death ligand that binds to or connects to the PD receptor on the T cell.

Inside Jason's body, malignant B cells had grown and were using PDL-1 to put the brakes on the killing part of his immune system. At the same time, now that the immune system had received a message that the cancer was "self" and not alien, the immune system actually set out to protect and support the cancer.

Dr. Lesokhin said it appears that "the tumor co-opts the immune system and says, 'I'm okay. I just want you to help me grow.'"

It's tempting to anthropomorphize the cancer and think of it as cunning, or strategic, but, really, cancer is a product of the same evolutionary processes that lead to our own survival, or that of any other species or organisms. When a mutation occurs inside us, it thrives if it has developed the ability to evade our body's defenses. Over the course of our lifetimes, we're being thrown tons of twists by malignant cells, and it takes only a handful to turn on the immune system's brake and start a cascade of malignancy.

"It's basically evolution at work in real time, a Darwinian survival system," Dr. Lesokhin said.

In the case of blood cancers, the exact mechanism is still being explored, but Dr. Lesokhin hypothesizes that the successful cancers result from an evolutionary process in which the surviving mutations are the ones that evolve a key adaptation that allows them to "use the immune system or avoid the immune system."

Jason had this growing inside him. A cancer had figured out how to turn down his defenses, mute them, while using the might of the immune system to build the infrastructure—the blood and tissue roadways and constructions—to help the cancer grow.

In Jason's immune system, there had been a coup. Left untreated, the malignant cells would've reproduced unchecked, voraciously eating up more territory, invading organs, causing normal bodily functions to slow or cease. Jason would've lasted only four months. Fortunately, there was a veritable nuclear bomb to deal with those rogue cells—or so it would have seemed.

Chemotherapy is brutal. "When you have cancer, you spread napalm on it and burn everything to the ground," Jason's oncologist, Dr. Mark Brunvand, told me.

Mostly through dumb luck, scientists had at least found an effective version of napalm for Hodgkin's, the kind of cancer Jason had. It provides a 90 percent survival rate.

The chemotherapy drugs target cells that are fast dividing, which is a marker of cancer. The malignant alien cells reproduce quickly, just like those healthy cells in a wound that are being fed by blood and protected by the immune system itself. The evil malignancies co-opt the system, and in an odd way, they get treated to the privilege of dividing quickly. There are other cells in the body that also divide quickly, including hair follicles and cells in the gut and mouth.

A fire hose was spraying Jason's Festival of Life with poison. That terrible toxic cocktail called ABVD was effective against all these cells, but its list of possible side effects reads like a who's who of extreme irritants and dangers: bruising, bleeding, tiredness, constipation, flulike symptoms, hair loss, mouth ulcers, sore eyes, dizziness, and on and on. On top of this is insomnia, which can be a byproduct less of the chemotherapy than of the use of steroids. These are used, as you know by now, to limit inflammation and cut down a massive immune system response. Why, you might ask, would you limit an immune response in a time of cancer?

In this case, you want toxins in your body. Poison is your ally, and the more it can be allowed to flow freely, the greater the chance it will target these fast-dividing cells. But part of the way that steroids suppress the immune system is by activating the adrenal glands (remember that when stress and adrenaline get activated, they suppress the immune system).

In short, there is nothing good about chemotherapy other than the fact that it can save your life. A trade-off, often, worth making.

Chemotherapy, Jason discovered, is also expensive. The first clinic he visited told him that the terrible toxic cocktail known as ABVD entailed "twelve chemos at $8,500 a pop. They realized I had bogus insurance, and they cut me loose."

He was two days away from treatment and needed a safety net. He found it at Denver General Hospital, a catchall for the uninsured or poorly insured, the place where you end up when they find you on the street with a gunshot wound or an opioid overdose, or where you go when you've got cancer and no money to treat it. It was October 2010. Jason jumped in, sort of. During his first round of chemo, he had trouble getting to his appointments on time.

He was "always on the road, always busy," said Dr. Michael McLaughlin, his first oncologist. "I thought: This guy is on the run."

Jason's chemo didn't work. He was one of the unlucky 10 percent whose cancers manage to survive the toxins. Sometimes this happens because cells mutate in the face of the drug's onslaught and become resistant to treatment. Relatedly, for the best chance of attaining good results, the treatment needs to be given in the proper doses at the right times. Jason thus didn't do himself any favors by missing some appointments, potentially giving the cancer more time to adjust to the treatment. Whatever the reason for the chemo's failing, and there can be no way of knowing for sure, the race to save Jason's life had begun.

37

Laughter and Tears

Jason could tell a yarn. He could hardly talk without telling a yarn; according to his zealous world view, every day was an adventure. He would relate his experiences like a combination of a bard, radio talk-show host, and bawdy comedian, punctuating his tales with occasional bursts of laughter, often directed at himself. His mother thought he'd missed his calling as a comedian—"the funniest person I know," she told me with understandable maternal bias—but often Jason's actions and audacity were the funniest part of the story.

When I think about Jason's cancer saga, I find myself thinking about a particular night of storytelling that took place by phone in late spring of 2011. It was a night when Jason and I began to reconnect in a much more real way, after having come to live very different lives.

I lived in a light-brown stucco flat in an almost-suburban residential neighborhood in San Francisco, and Jason lived in Las Vegas and his van. The night he called, I'd helped put our toddlers to bed—Milo was two and his newborn sister just six months. They slept in a back bedroom; Meredith, my wife and their mom, read in the room next to theirs. In the front room of the house, I sat on a giant blue ball we used to bounce the kids on when they had trouble calming down or sleeping.

Jason talked about his cancer. He discussed missing chemo

appointments with the same self-effacing humor he'd employed when he talked about forgetting to study for a French test in eleventh grade. It was nothing to get worked up about, or even a badge of honor, especially when there were so many more adventurous things to do.

He launched into a story about driving across the country to a sales conference. By this time, he had the Ford Windstar. During the trek, he'd wound up driving through Kansas, where he'd heard on the radio there was a good high school basketball tournament going on, and he'd decided to make a pass through and watch the games.

"The motels were filled up," he said. "I slept in the van. The van was full of trinket boxes, and it was so stuffed in there that I could barely make room on top of them. I could barely fucking breathe!" I feared his happy laughter would wake the children. I was right there with him, totally swept up in his journey.

Then *boom*, he'd be off to the next topic. He told me off-color stories of past dating conquests and failures, and he also suggested those days might well be behind him. "Dude, have I told you about Beth? She's awesome."

Beth Schwartz, his girlfriend, had *angel* written all over her. She was right in Jason's wheelhouse. She loved football—she'd been sports editor of her high school paper in Houston; she was an athlete herself, a soccer player and track runner; and she loved to laugh and thought he was hilarious. He found her beautiful. He also might not have fully grasped just how wide the latitude of her ability to either ignore, or even appreciate, his flights and dreams.

They met Labor Day weekend in 2006. She'd broken her leg in an in-line skating accident and was on crutches when she showed up at a West Virginia alumni club mixer to watch the Mountaineers play on the screens at Sierra Gold, a medium-size tavern-style joint. Beth was in a back room when she overheard someone say her name. It was some old-timer in the alumni club, who answered "Beth," in response to Jason, who had asked him: "Who is that girl on crutches?"

Jason was in the bar for work purposes. He was setting up a fantasy football terminal that he'd come up with. (Beth called it "a crazy stupid Boondoggle Sports Network.")

She told me: "I looked at him and thought: 'Uh-oh, I'm in trouble.'"

Why, Beth?

"He just looked like trouble—somebody in Crocs and cargo pants and a T-shirt that looked like it hadn't been washed in a few days."

Soon they were drinking at the bar. While Jason was trying to seal the deal, another man tried to hit on Beth. The guy had tried to make a joke about how Beth looked so young, and he wondered if her mother was at the bar too. It was awkward, and Jason smoothly retorted: "Let me give you some friendly advice. Never try to pick up a lady by asking about her mother."

"He had me. Right there," Beth said.

In addition to Beth's other attributes, she had a job that played into Jason's love of adventure. She was editor of a high-end luxury magazine in Vegas, which meant she got invited to every notable restaurant opening and concert. It was Las Vegas, on the house. "Being the Italian talker he was, he could sit there and hold court—if I could get him to dress properly," Beth said.

There were quiet times too. A typical date involved the pair of them going to a bookstore or a coffee shop to read. Jason devoured history books, and loved that Beth was also an avid reader. There were times when it could feel downright domestic.

That night, on the phone, Jason had a question for me.

"Rick," he said (short for Richtel, and what he often called me), "do you think I should have a family?"

I listened to his question. I couldn't tell if he was serious or not.

"You, Noel, Meier, everyone has settled down, and you guys seem really happy. I've been thinking about whether I'm running out of time." He sounded mournful.

"It's great, Greenie. You'll love going to bed at nine." I was half joking, but also trying to soft-shoe the conversation to feel him out.

"I'm serious. Should I?"

"I'll tell you one thing. It's freeing as hell. I spend a lot more time thinking about stuff I love thinking about, writing and playing tennis and even playing music, rather than where my next date is coming from. And having kids and a wife you love—well, it's impossible to describe how great it is until you're in it."

"I don't know, man . . ."

He really did hate this subject; that much I learned as time went by. He loved Beth, cherished her, but when I asked her if he wanted to put a ring on it, he'd shut down the conversation. It had nothing to do with Beth, I realized, and a lot more to do with commitment. Maybe because of the loss of his dad or his love of the open road—I could never be sure.

It dawned on me that night, or shortly after, that my own relationship with Jason had transformed. We were genuine friends now, and a lot of that had to do with my own bout with illness. I told Jason what I'd been through.

It happened when I was twenty-five.

I can picture the moment, if not the day and precise year. I'll put it around late 1991 or 1992. I was out for a jog in Palo Alto, where I worked at my first newspaper job. I felt light-headed. This had been happening off and on. I went to a doctor who'd been assigned to me by my medical insurer. He was a great guy, in his seventies or early eighties, I'd guess. I'd gone to him a few times to relate these symptoms, and he just gave me antibiotics and kindly sent me on my way.

Even I knew it was not the right course of action. I was due for a self-reckoning.

About three years prior, after I graduated from UC Berkeley, I went to Europe with friends. While in a youth hostel in Rome, I wrote a life-changing postcard. It was written midsummer, to the School of Journalism at Columbia University, which I'd applied to and which had put me on its waiting list. The postcard rhymed and explained that if they didn't take me off the waiting list and admit me, I would spend all my tuition money on booze.

I had little business being on the waiting list at the esteemed school in the first place. I'd never done any journalism, which tended to be a prerequisite. The reason I'd applied, in my final semester at Berkeley, was driven by a gut-level understanding that I liked to write and ask questions and explore ideas, that I had something of a heightened curiosity instinct. True story: Two days after I returned from Europe, I was back in Boulder, with no freaking clue about what I was going to do with my life, when the phone rang.

"Is Matthew there?"

"This is he?"

The guy introduced himself as the assistant to the dean of Columbia School of Journalism.

"School started yesterday and a spot opened up. I'll be honest, Matthew, you were far down on the waiting list. But the dean saw your postcard and thinks you're funny. Would you like to come to Columbia?"

Pause to make sure it wasn't a high school buddy playing a prank.

Sure, yeah.

Fuck.

At Columbia, I exhibited a devil-may-care exterior. Outwardly, I still thought of myself as belonging to that too-cool-for-school club Jason had christened the Concerned Fellows League. But I was petrified, the youngest, most inexperienced member of the class. The terror went deeper than that. As I look back, I realize that this was the moment when I subconsciously decided it was time for me to become something great, whatever that meant. My childhood aspirations had come home to roost. I could be the Jason of journalists. The thing is, a sizable part of that emotion wasn't authentic interest—I didn't yet understand what it means to be a journalist or a writer. I just knew I wanted to succeed. This terrible disconnect helped explain the fear; I knew, on some level, I had only generic goals, not ones true to me.

Why tell you all this?

Because it explains why I stopped sleeping. I don't mean just restless nights. I hardly slept at all. I would go full weeks, largely tossing and turning, sleeping only a few hours at a time, trying to figure out how to tackle stories I didn't understand or attack classes that I wasn't sure I found interesting. Or maintain a facade of calm that had no existence in reality.

After I realized something was wrong, I spent three intense years learning a lot about myself and this sickness, and that's exactly what it is—a disorder. In practical terms, its toll on my behaviors ranged from spending days sweating from exhaustion to struggling to focus on work to making stupid social choices and, more than any of that, to setting the stage for severe anxiety and depression, letting my adrenaline run wild to keep me functioning without good rest. When I took on this book, I went back and learned that what overwhelmed me then had very much to do with the immune system and its relationship to sleep and stress, even though it looks like a matter of "mere" psychology. There was some of that too.

In the course of researching this book, I described my situation at that time to Dr. William Malarkey, a professor emeritus at Ohio State University and an expert in how the body's stress and neurological systems relate to immune function. He's worked closely with Janice Kiecolt-Glaser and Ron Glaser, and is an expert in the causes and impacts of stress.

"You were going through a search for a mission and meaning," he said. "At some point, you took this incredible long shot"—meaning applying to Columbia—"that you hadn't thought out. You happened to hit a home run. Suddenly you think, 'I have to be Babe Ruth. Now I'm among major league players.'"

Then he switched to biological terms. He said that fight or flight took over as if "I'd been thrown into a lion's den, or with a pack of bears."

This, he said, was obviously not true. But that's how I perceived it, and I and many others tend to make the same mistake. "What happens today is that many people are living with imaginary bears at every step of their lives—something in the news or around the bend is going to get them." What followed was what he called a "norepinephrine high."

It was a survival mechanism, in the short term. But in the long term, it was dangerous, even deadly.

As I discussed earlier, norepinephrine is one of two major neurotransmitters or hormones—a signal secreted from the nerve endings or the adrenal glands—that are released immediately as part of a fight-or-flight response. The other major hormone is called epinephrine, or adrenaline. When we perceive a dangerous situation or any kind of threat, these get secreted and start to impact other cells in the body. "You get thrown into the lion's den or surrounded by bears, and you begin to be alert to everything that's going on around you."

Immune cells are among those impacted. In fact, according to Dr. Malarkey, the connection between the immune system and the adrenal system may be so intimate that it's difficult to separate the two.

I told Dr. Malarkey that norepinephrine and epinephrine sound much like interleukins in that they are sending signals that impact immune cells. He just laughed. "Exactly!" he said. "I've been saying it for years. The difference is they were discovered by people in different fields. If they'd been discovered by immunologists, they would've been called IL-1 or IL-6 or something."

Leaving semantics and returning to substance, he said that norepinephrine and epinephrine can start to feel, perversely, exciting. "You get addicted to it. You need it. Suddenly it's going on all the time. The brain is driving it. Now you get all the disruptions of excess. You get dis-regulation of the immune system."

Dr. Irwin, the sleep expert from UCLA, explained that what followed was "a sickness syndrome, sickness behaviors, driven by inflammation." Feelings of depression, social isolation, withdrawal, fatigue.

That's exactly what happened next.

For that period in the mid- to late 1990s, I battled to discover myself. I realize this is an overused phrase, somewhat trite. Here I'll defend it as central to health. I wasn't going to stop until I had a better understanding of what was consistent with me. I had long moved past questions like what I wanted to be. Quickly, that stupid idea dissolved into much more basic questions: What was comfortable to me? What activities and circumstances felt right?

The need to answer those questions was amplified greatly by insomnia. The level of sympathetic response that I was experiencing on a daily basis, still having trouble sleeping, was clearly impacting my health, my well-being, my level of anxiety. I can fairly say that I was experiencing a kind of addiction to that adrenaline, the norepinephrine and epinephrine. It felt like excitement. But it was betrayal.

I solved it by backing into the science. I started meditating. I can't remember how or why, beyond the obvious idea that the concept was out there.

I can still picture one night when I was lying in bed, breathing deeply, and I kept meditating. An hour, more than an hour. I felt my jaw go slack. I felt my body calm. I fell asleep. I woke up in the morning rested—genuinely rested. Feeling different than I had in a long, long time. I kept doing it. Many nights, it would take an hour or more, maybe two.

Now that I have learned the science, I know I was shutting off my sympathetic immune system. I was short-circuiting the dangerous cycle that Dr. Irwin had described in which my central nervous system had dosed my body with adrenaline, further intensifying the fight-or-flight response, inflaming swelling and the immune system, leading to further adrenal responses. I don't know what toll that period took on my longevity, but I wouldn't trade anything for what it taught me.

Simultaneously, I'd drained my psychological tank dry. Sat on a shrink's couch and sobbed. I accrued an unpayable debt to my parents, my girlfriend, and a buddy I'd met in journalism school at Columbia named Bob Tedeschi, who had become a brother. I mention these people not for the sake of merely giving thanks, but because the science bears out that the ability to find connection during times of sickness, including anxiety and depression, is instrumental in healing. It helps the immune system find balance, and this makes sense from an evolutionary perspective; the idea that you are part of a community is powerful incentive and motivation for your body's machinery to seek harmony. Left alone, you might recede further.

During this period, I scoured every corner of my brain and began to realize the wisdom of the old adage that there is little to fear except fear itself. Looking back, I can see a close association between the end of my psychological probing and the beginning of my period of meditation and relaxation. I had, in plain terms, given myself permission to relax. I'd become somewhat comfortable, and then eventually, I had nothing to prove. I'd learned, the hard way, to let myself listen to me, first and foremost, and to suppress the voices of others.

It is also impossible to overstate, as it speaks to my health and, I suspect, the health of many. I emerged as someone with a sense of confidence to trust myself, which in turn allowed me to listen to the parts of life that excited and motivated me, the kinds of environments and friends that made me comfortable, and the things that I needed to shed as inauthentic. I'd found self.

The best example I can offer about the value to health of hearing and following my own voice, and not chasing generic external validation, took place for me in the late 1990s. I was freelancing for the *New York Times*. It was going well. I loved the aspects of journalism that I'd suspected I would all along, writing and exploring and being curious. But I also was thrilled to be freelancing. I was working more on my own terms. This wasn't about the grades anymore, or the approval of the boss. I worked and I liked it and I got paid and I'd stopped aspiring to rise in any ranks.

Then the *New York Times* offered me a job. It was a dream opportunity for a still-young journalist. The only catch was that I had to move to New York instead of remaining in San Francisco, where I had moved after college.

The idea scared the hell out of me. I knew in my heart that I didn't belong in that city, where I'd suffered at Columbia, and in an environment where I feared I might lose track of my priorities—days spent in a hypercompetitive world with the feeling of people looking over my shoulder. I pictured myself getting caught in the spiral of adrenaline, enduring long days in an office with people more able to sustain that than I was, or more willing to. I turned down the job.

Remarkably, the *Times* relented. They'd hire me, they said, and I could stay in San Francisco. Two years later, the paper changed its mind. I was told I'd have to move back to New York. "That's what everyone does," an editor told me. It was nothing personal.

I flew back to New York and made my case to be allowed to stay in San Francisco. I told them they seemed happy and I was happy and nothing was broken. An editor told me: "This isn't about being happy. It's about doing the drill everyone has done."

That idea, I'd discovered, was anathema to me.

I was given a date: October 1, 2001—move to New York or be fired. By then I'd started dating Meredith (not Merredith Branscombe, but another fellow Coloradan named Meredith Barad), the amazing woman who would become my wife. On October 1, I woke up and went to my desk and waited for the phone to ring. I

worked and I waited. The next call, any call, could be the one telling me was I fired.

It didn't come, not for a week, then not for a month, and then
not for more months. I kept writing and living and doing my
thing, and more and more I trusted my voice, my muse. I married
Meredith, and I started writing books, and they were foremost for
me, stories I told myself, that poured out, all kinds of songs, using a
voice so far from the one I'd parroted for a long time. One day,
the *Times* relented. They were happy and I was happy.

Again, this is not an aside. And, again, the value of that cannot
be overstated as a lesson of the immune system. The more consistent I became with myself, the more I jettisoned what was alien,
the healthier I got. I also tell this story because it allowed Jason
and me to become real friends, on a much more honest footing
than we'd shared as kids.

Jason, meantime, had been following his own muse. It led him
to come up with one business idea after the next, selling, schmooz-
ing, spinning yarns of the power of the gadgets and ideas he truly
believed were new and different—from mobile phone minutes to
power juice blenders and on and on until he'd settled on his latest
venture, casino trinkets.

That night in the spring of 2011, the contours of a new relationship
took shape. We talked about life, and we talked about cancer.

"I need to beat this thing," Jason told me, "and then I can figure
out what to do next."

"So how's that going?" Cancer.

"I'm not going to lie to you, Rick, it fucking sucks."

He told me about the chemo, and how it wrecked his body, and about how he had to take a steroid to keep the inflammation down, and that kept him up at night. "It's the sickest I've ever felt, times a thousand. I just sit there, starving for sleep. I hurt all over and I'm lying there and I can't even read or watch TV. It's brutal. I wouldn't wish this on my worst enemy."

I guess I'd have wanted to miss a few chemo appointments too.

Jason had failed his first treatment. There were other options in the chemotherapy realm and he was headed for them.

In the background, though, a new field of medicine was taking shape that built on all those years of hard-hat immunology work. It was called immunotherapy. The science behind it blows the mind.

38

The Lazarus Mouse

A substantial part of the immune system revolves around the way information is communicated. Molecules send and receive signals, urging immune cells to attack, do further surveillance, withdraw, implode, lurk, help new tissue grow. In the broadest sense, this information is transmitted in two different formats, or media.

Some of the communications are known as soluble, or fluid-like, and involve interleukins. These molecules are released and can travel around and infuse other cells with instructions.

The second type, which I've already also described and will elaborate on here, involves molecules or proteins that appear on the surface of cells and that connect, or bind, to a molecule or protein on another cell. These are like antibodies. They travel the body not in fluid form but attached to a cell, and then they connect to another cell at an extremely specific spot. These are mates in a puzzle that require physical proximity.

The concept is important because it helped save Jason's life. To show how, I need to delve a little more deeply into the science.

Typically, one piece of the puzzle is called a ligand, and the other is a receptor. (Ligand comes from the Latin *ligare*, meaning to bind.) A ligand binds to a receptor.

Through the 1980s and the 1990s, immunologists did a lot of looking for molecules on the surface of immune cells—basic archaeology—and then tried to find their mates. One reason they would hunt for these pairs was in hopes of finding a match that would help explain what each molecule was doing on the cell surface in the first place. What did the individual pieces of the puzzle do, and what happened when you put them together?

"Every piece builds a story. It's like getting to know a friend. Same thing through a series of encounters through molecules," said Matthew "Max" Krummel, an immunologist who was there at the moment of one of the twentieth century's major scientific moments—when CD80 and CD86 met their mates.

This is the story.

In the late 1980s, work had been done that identified two ligands that are expressed on the surface of two major immune system cells, B cells and dendritic cells. Scientists had discovered that these ligands bind to specific molecules on the surface of T cells.

As these various immune cells circulated in the festival of our lives, they would bump into one another. If the surface of a B cell had the right ligand and the surface of the T cell had the right receptor, the two molecules would bind to each other. This would set off a reaction.

Okay, fine, but so what? What was the reaction?

So we're trying to cure cancer here, give it a sec, will you? Stick with me.

In plainer English still:

T cells can attack invaders and organize attacks. Researchers discovered molecules on the surface of T cells that connect to molecules from other parts of the immune system, namely B cells and dendritic cells. In other words, scientists had found pieces of a puzzle that fit together but without knowing what the puzzle looked like—or exactly what that puzzle meant. They discovered that one of the key molecules on the surface of the T cell was called CTLA-4. Another was CD28.

One more bit of deep trivia that is not at all trivial: CTLA-4 and CD28 both bind to ligands called B7-1 and B7-2—also known as CD80 and CD86.

Okay, so then what?

Around 1989, CTLA-4 was being explored jointly by two eventual all-star academics in the immune sciences, James Allison, who was at Berkeley at the time, and Jeffrey Bluestone, who was at the University of Chicago and eventually the University of California at San Francisco. There was a third researcher named Peter Linsley, from Bristol-Myers Squibb, a pharmaceutical company, doing related work.

Bluestone and Allison weren't particularly interested in cancer, or rather, that wasn't their central focus. They were concerned with the immune system overall.

At Berkeley, a PhD student in Allison's lab performed an experiment that entailed taking a tumor from a mouse, putting it into a test tube, and then injecting it with foreign genes. The impact of injecting these genes was to cause the tumor cells to present the molecule called B7-1, the ligand that binds to receptors on T cells called CTLA-4 and CD28. The researchers then injected T cells into the test tube, and lo and behold, the T cells attacked the tumor in force, attracted by B7-1, and they wiped out the malignancy.

One more time, gently, for good measure. Researchers figured out how to display a puzzle piece that attracted another puzzle piece and this stimulated an immune system response to wipe out a tumor.

Good news, right?

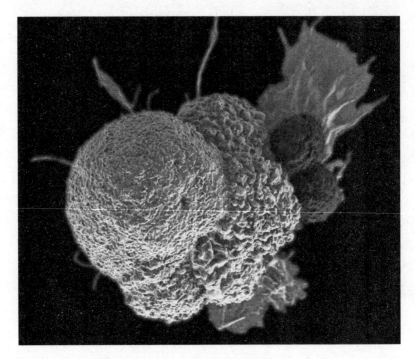

Two T cells (right) attack a cancer cell. (NCI/NIH)

Yes, a huge step in the right direction. But this wasn't yet the Holy Grail. The steps were too artificial, like tailor-making a tumor by putting foreign genes into it so that it could be targeted. Plus, this whole thing had happened in a test tube. This wasn't yet a solution to allow manipulation of the human immune system. But it was a powerful indication that such a solution was possible.

This is where Krummel began his collaboration with Allison, who won the 2018 Nobel Prize for what happened next.

Allison and Krummel decided to experiment further with CTLA-4, the other molecule that had bound with B7-1 and B7-2. They soon noticed a curious thing. When CTLA-4 was attracted to and bound to a ligand, the immune system didn't ramp up as it had in the mouse experiment. Instead, the immune system seemed to be dampened or to have no effect at all.

"I thought, we gotta figure out what CTLA-4 does," Allison reflected. Something about it nagged at him.

Krummel and Allison asked a question: If CD28 causes T cells to multiply, but CTLA-4 seems to have no effect, what might happen if you combined these agents?

What they discovered was a turning point. Stimulating CD28 led to an increase in T cells, and a heightened immune response. But when CTLA-4 was mixed in, it brought down the level of T cell response. Not only that: The more CTLA-4 was added, the fewer T cells proliferated. That suggested that CTLA-4, rather than causing the immune system response to increase, was causing it to turn down or even off.

They sensed they were on to something big.

Krummel devised a chemical process that would allow him to create varying levels of CD28 and CTLA-4 such that he could begin fine-tuning the amount of T cells created. The year was 1994.

"We could turn T cells up and down like turning a stereo up and down," Krummel said. Or, if you prefer a different metaphor: "We found a hot tap and a cold tap. We immediately had this whiteboard discussion," he said. What did this mean and what could they do with it?

They started experimenting, trying all kinds of combinations, one after the next. "In the course of nine months, we went from volume control—hot and cold—to every single animal model I could touch, pushing T cells to grow faster, watch them grow slower. That's when Jim brought in the tumor model."

Allison, by now steeped in this as virtually no other scientist, turned ideas over and over in his mind, trying to make sense of it all. What did these molecular interactions add up to? He joked with me about the pieces finally coming together one night in 1994, while his mind was wandering after "too much wine." He thought he might understand how cancer was playing a trick on the immune system, allowing the disease to evade our defenses. And he had an idea how to reverse the trick.

Allison had invited a postdoc named Dana Leach into the lab. Leach brought the rodents with the tumors, which now were out of the test tube and into actual critters. The vet injected rats with several fast-growing cancers. The researchers let the cancers blossom. Then they injected the mice with a molecule—an antibody—that was aimed at disrupting any connection that the cancer cells might be trying to make to the CTLA-4.

The idea was to see if they could keep the cancer from turning on the brakes of the immune system by disrupting the communication between the cancer and the immune system.

"We were just trying things out," Krummel said.

A few days later, Allison came in to check out the progress. "I went, 'Holy shit! It cured all the mice.'"

The previous experiment had entailed isolating the tumor tissue in a test tube and then modifying its genetics so that it would stimulate a T cell response. This was ultimately impractical.

But in the new experiment, the researchers did nothing at all to the tumor. It was just a tumor, like one that might be growing inside any of us, eventually inside Jason. It was in its natural state.

Instead of changing the tumor, the researchers added antibodies to interrupt cancer's trick and stimulate a response from the immune system. Specifically, they inserted an antibody that bound to the immune system so that it would take off the brakes in our elegant defense.

"What was surprising is that we hadn't given the immune system any new information about the tumor," Krummel said. "There was a set of preexisting cells"—the T cells—"and they were raring to go."

When Allison looks back, he thinks of the immune system in a vastly different way than we conceived of it for the longest time. He doesn't think of it merely as a powerful killing machine, not at all. Instead, he thinks of it as sharing killing powers with extraordinary powers of restraint. One of the chief jobs of the immune system is to shut down its attacks, hit the off button. Screeching brakes get applied to the T cells.

"They get a signal. They kill themselves. If it didn't work, people would get diabetes, multiple sclerosis, lupus," he said. "By far this negative selection is the central tolerance, to get rid of T cells; 90 percent of every T cell that's developed gets killed."

He'd figured out what CTLA-4 did. "CTLA-4 is there to protect you from being killed by your immune system."

Whoa.

But isn't cancer killing people? Why would our bodies allow the brakes to come on in the face of a deadly tumor?

The answer is related to a trade-off with wound healing, which is one of the most important functions of both the body and the immune system.

39

Wound Healing

If you get an acute injury—say, you step on a stick or cut your hand on the edge of a can—the event sets off an urgent survival cascade. Red blood cells rush to the scene and begin to coagulate or clot. They stop the bleeding. Cells from elsewhere move into the gap and start to divide. These include immune system cells, neutrophils and macrophages.

Sabine Werner, an expert in wound healing, describes it like emergency crews arriving on the scene. "There are very rapid events to get closure through the blood clot." The immune cells are there to deal with "bacteria, fungi, viruses—they can all be there."

The neutrophils produce proteases, which attentive readers now know of as enzymes. These enzymes are a bit like a grenade. They make holes in certain bacteria, "killing them actively. Bacteria are also eaten by neutrophils and by the macrophages," Werner said. Tidy.

In addition to neutrophils, there comes a second vicious killer. It has one of those impossible-to-remember names: reactive oxygen species (ROS).

Just remember this: It's nasty. One such reactive oxygen species is hydrogen peroxide. The macrophages and neutrophils both can synthesize the chemical to kill in the area of the wound. The neutrophils and other killers didn't just take out bacteria or other possible infections. They also killed some of the other surrounding tissue. This is the reason that often, after you experience a wound, even a minor one, the pain and inflammation are worse in the days that follow the event. Your immune system has done housecleaning with industrial-strength chemicals.

The area has been cleansed of "other," leaving scorched earth. Then in the dead zone, the macrophages eat.

Almost as quickly, construction workers move in. In the early 1990s, Werner, researching the phenomena, noticed that a wound site, within about one to two days, exhibited a tenfold increase in growth-promoting signals. She began to vigorously pursue the question of what is happening within the body that allows such rapid healing. Where are the signals coming from that allow cells to divide quickly and replenish the tissue?

Think about how dramatic this transition is. One moment your tuna-can-slashed finger is being swept by a SWAT team, and then, within hours, an entire construction process has come in and replaced the killing machine. And what were the implications for the overall system of health? "I got really excited at how fast a wound can react to this," she reflected.

She didn't know yet about the dark side.

The reconstruction process has, of course, its own complicated language. The term for one of the key cell types stimulating regeneration of our tissue is *fibroblast*—highly versatile and hearty cells that proliferate and migrate to the site. These cells are drawn by signals sent by macrophages. This is of note in that it shows a different side of the macrophages. These "big eaters" also play a role in stimulating the growth of new tissue.

As the fibroblast cells come together, they form connective

tissue, a bridge between the new and old tissue. At the wound site, the new tissue takes on a granular quality, hence its name granulation tissue. Crucially, these tissues are fed by blood vessels that spring up around the edge of the wound, creating veritable feeding tubes for new tissue. A kind of tenacious web forms, a fibrous matrix that, as Werner and her coauthors put it in one paper, protects against invading pathogens and "is also a reservoir of growth factors that are required during the later stages of healing and it provides a scaffold for the different cell types that are attracted to the wound site."

In the Festival of Life, a particular party spot is imploded, then cleared away of debris. Next comes the construction of the foundation and of scaffolding, and then rebuilding starts. But as is true of many construction projects, permits must be obtained. The body must accept that what is being built is approved of as "self." Anything seen as alien to the point of being pathogenic will be destroyed, and the site will not be rebuilt.

There is a dangerous corollary. Once permission is given, once the new cells being nourished are deemed "self," the construction can go on with zeal. The trouble is, the new cells aren't always self. Sometimes they are cancer.

And so the factors that promote growth of healthy tissue also appeared to promote the growth of tumors. This was an idea that had been floated since 1863, when Rudolf Ludwig Carl Virchow, a German scientist, observed: "Chronic irritation and previous injuries are a precondition of tumorigenesis."

Werner gives talks in which she cites two other equally prescient quotes:

"Tumor production is possible overhealing," commented Scottish physician Sir Alexander Haddow in 1972.

And then there was the observation of Harold Dvorak, a pathologist in Massachusetts, who in 1986 said, "Tumors are wounds that do not heal."

The wisdom of these statements has been borne out in powerful lab experiments.

One decades-old telling experiment had been performed using baby chicks. The experiment, done at Berkeley, involved injecting into chickens a virus known to give them cancer. The chickens were injected under the skin or in muscle. In either case, the injection caused a minor wound.

Within one to two weeks, a tumor appeared, usually at the injection site. The chicks died within a month.

The researchers decided it was reasonable to assume that the wound itself was relevant to the growth of the tumor and came up with a second experiment to prove that point. This time, they infected a chick in the right wing but not the left wing. At the same time, though, they pierced the left wing. Lo and behold, a tumor formed at the injection site and at the site of the wound on the left wing. The tumor on the wing that was wounded but not injected took about 20 percent longer to appear.

Something about the wound clearly was playing a role in promoting the tumor.

In the 1990s, Werner started to put the pieces together. What she and others discovered begins to explain why things like smoking or coal mining or sunbathing are so carcinogenic. Each activity injures the tissue and damages the DNA. When the tissue is damaged, the immune system kicks in and cleanses the site and helps stimulate new tissue growth. The trouble is that when the

DNA is damaged, the new cells that grow can be malignant cells, some that are made up of self but that are different enough to behave like a cancer. These cells aren't playing by the normal rules of the body and staying within their boundaries. Add all this together and you can wind up with cancerous cells that are protected and even nurtured by the immune system.

This explains too some risks of cancer experienced by sufferers of certain autoimmune disorders that cause chronic tissue injury.

When a wound occurs—an insult, as it is known in scientific circles—cells divide. Of course they do. New tissue is needed. But when new cells divide, there is always a chance something can go wrong. Each cell division is an opportunity for a mistake, a mutation. A piece of DNA might get incorrectly copied, for instance. This happens all the time. Fortunately, in most cases, this mutation has no consequence because the cell dies or gets rapidly devoured. The mutation is so unusual that the cell can't survive because essentially it lacks the genetic material to live, and the macrophages eat the refuse. Story over. At other times, the mutation is picked up by the immune system as being sufficiently foreign so as to be potentially problematic. It is bombed or blown up, destroyed and then eaten. Story over.

Sometimes, though, the mutation is extremely subtle. The cell has the genetic material sufficient to survive, and it is sufficiently like "self" that it isn't recognized as problematic by the immune system. In some cases, the immune system tests the material but decides that it is more likely self than not and leaves it alone.

This doesn't mean that such a cell is necessarily cancerous. A cell with a single mutation is highly unlikely to be cancer. Werner explained to me that a cell that turns cancerous needs to undergo at least five to ten different genetic changes. Not just that, to be a "perfect bad cancer cell," the random genetic occurrence needs specific changes in different regions of its DNA. For instance, a mutated cell that is likely to survive and become cancer has chanced upon the ability to send signals to immune cells with the instruction: Don't attack me; protect me and nurture me.

"They secrete factors which change the immune cells," Werner told me. For example, "the macrophages are not inflammatory anymore, but rather, they protect the cancer cells and stimulate the formation of blood vessels."

This is a juncture where cancer takes advantage of the immune system. The cancer grows and grows, quietly protected, fed by blood vessels, even guarded by fibrous networks. The tumor "is cruising along, invisible and growing," said Allison, the pioneering researcher of CTLA-4.

But then "at some point, [tumors] reach a certain size and they can't get enough oxygen, enough food," Allison explained; they have gotten too big for their environment. "They start dying," and the macrophages come in, phagocytosis takes place, the tumor debris gets cleaned up, and then the immune system starts providing more growth infrastructure, as it would in a healing wound, and simultaneously CTLA-4 shuts down the attack dogs.

It's a vicious cycle perpetuated by the immune system. Full stop. *The immune system starts to feed and nurture the cancer.* Your elegant defense has turned against you.

What this adds up to is that the likelihood of getting cancer depends in large part on how often a person experiences injury or certain types of injury. This is just math. More injury means more cell division and, simply, more opportunity for dangerous mutation to occur.

Enter one of the world's biggest killers.

When someone smokes a cigarette, tiny little wounds are created inside the fragile pink tissue of the lung. Into the lungs pour several thousand chemicals, including a number of them that not only damage the DNA but that interfere with repair of DNA. Meanwhile, the police and fire brigade of the immune system shows up, and the process of wound healing begins. New cells are created. Over and over and over, cigarette after cigarette, year after year. (Smoking is a chronic activity, as opposed to, say, inhaling fewer chemicals less directly at the occasional campfire.) In the case of smoking, the malignant cells are fed and protected by the same system that cleaned out the wound in the first place and made sure that no pathogens were there to inflict harm.

Some of those new cells are mistakes and are recognized as non-self. And some have the right combination of random mutations to live and look a lot like self, so much so that the immune system, the very system created to defend us, becomes the promoter and protector of the tumor.

Again, in key respects, cancer is just a numbers game. The more wounds you get, the more mutations and inflammatory events, the greater the likelihood of cancer. That's what makes things like smoking so potent. The risks grow with each puff. Similarly, sun exposure, absent sunscreen, presents another opportunity for a wound and an inflammatory response, which, combined with mutations directly induced by UV irradiation, enhance the risk of the development of skin cancer, including the particularly dangerous melanoma. Other toxins that come into the body, whether food toxins or chemical ones, can also create wounds, places of insult, even minor, that require repair, inflammation, rebuilding. Each minor assault is a chance for cell division and an immune system response that, while intended to cleanse, might also lead to cancer. Smokers are almost certain to get cancer at some point, due to the very certitude of math. If you're a smoker, you might have cancer right now. In fact, you *probably* have cancer right now. Most likely, though, it lacks the precise types of genetic changes that will let it proliferate by, in particular, co-opting the immune system. Just because cancer exists doesn't mean it will take hold.

Those of us who do not engage in such high-risk behaviors are much less likely to get cancer, or rather, we're not as likely to get it as quickly. But if we live long enough, math will catch up with us too.

The fact that you're likely to get cancer eventually, and that it will take hold *eventually*, says a mouthful about the trade-offs being made by your immune system. It has evolved so that it allows for the possibility, even likelihood, of cancer taking root. The reason is simple: It is willing, in the short term, to risk a mutation taking hold in order to allow for the immediate rebuilding of tissue. What, after all, would be the alternative? Leave holes in your tissue? Allow your body to be chipped away at, one nick and cut at a time?

Cell division is a must. Mutation, cancer, is a by-product of cell division. It is one reason your death is preordained. This dynamic, though, also holds the keys to combating cancer. That's what Allison began to exploit with CTLA-4. A second major conceptual discovery has helped tinker with the immune system to turn the odds in favor of life.

40

Programmed Death

Recall that James Allison had discovered we could modulate the immune system by playing with CTLA-4. That's the molecule on T cells that helped dampen or kill an immune system response.

What Allison and Krummel and others in the lab revealed was that tumors appeared to stymie the immune system by taking advantage of this molecule that is crucial for survival. The tumor stimulated a braking system that keeps our elegant defense from going berserk and overheating, causing inflammation, fever, auto-immune disorder, and so on. But the cancer in mice, the Berkeley scientists discovered, was sending a signal to activate CTLA-4 and thus cause the immune system to come to a standstill. In this way, the cancer could grow unchecked by the immune system.

CTLA-4 turns out not to be the only such brake. One of the others is known as PD-1. The PD stands for programmed death, which I've briefly described already. It is a molecule on a T cell that causes the immune system to self-destruct—in effect, to commit suicide.

The notion seems incredible on its face. But it's very common. The discovery was made in 1992 in Japan by Dr. Tasuku Honjo at the Kyoto University School of Medicine. He hadn't been looking for a discovery nearly as profound as the one he made. So profound was the work that Dr. Honjo would share the Nobel with Allison in 2018. Dr. Honjo and his team had been trying to understand what the Cancer Research Institute described as a "normal cellular housekeeping." The researchers scoured genetic material until they found what looked like a gene involved in prompting some cells to die when they were no longer valuable. This was called programmed death, a kind of suicide by cells no longer useful to the body. Dr. Honjo and his team went deeper into the origins and function of programmed death and found that when they disrupted or knocked out the PD-1 gene in mice, a huge portion of the rodents developed autoimmune disorders approximating lupus.

In other words, the programmed death gene appeared to be involved in suppressing the immune function.

Why would it make sense for an immune cell to commit suicide? For the same reason that there are so many brakes in our body's defense network. It's one more fail-safe process, another way to keep the most powerful, free-ranging system in our body from going rogue.

Across the Pacific Ocean from Dr. Honjo's lab, in Silicon Valley, the initial discovery of programmed death was met with great interest by Nils Lonberg, a scientist and entrepreneur who thought he could use it to cure cancer. He'd been planning for this moment for years, ever since he started milking mice.

Lonberg, born in 1956 in Berkeley to a chemist father and a psychologist mother, began his own pioneering cancer work indirectly, with a dream of making transgenic mice. This involves genetically engineering mice to harbor human genes. Off topic as this might sound, it happens to be directly in line with the field of immunology, going back even before Jacques Miller was toying with mice to discover the role of the thymus, and without exaggeration, it extends forward to saving Jason's life.

By the mid-1980s, technology had moved far beyond Jacques

Miller's shed. By this time, the idea was to use sophisticated genetic techniques to create mice that were mice, mostly, but with key human DNA spliced in. That way, it would be possible to see the effects of a particular molecule or drug on human DNA without killing a human subject.

But putting human DNA into a mouse isn't easy, or it wasn't at the time—"crude and physical," Lonberg described it to me. He was at Memorial Sloan Kettering in New York. He'd mate two mice around midnight. Then early in the morning, he'd take embryos from the female, inject the human DNA he wanted to embed, and then transfer the embryos into a "pseudo-pregnant mouse," meaning one that was primed to give birth. "Three weeks later, you'd get pups, little baby mice, with human DNA," Lonberg said. Then, through subsequent breeding, you'd get a purer form of the DNA-mouse.

(As an aside, Lonberg was in the lab one night, milking a mouse, when his wife entered. She is a scientist too. "She walked in. I had a mouse hooked up to the vacuum, milking it. She just looked at me," he said, laughing.)

Lonberg figured that if you could make a mouse with fully human DNA, could you make a mouse with human antibodies? If so, what would you do with those antibodies? Could you turn a mouse into a factory for building specific molecules of the human immune system?

If you could, then just maybe you could inject those antibodies into a human to support the person's immune system without risking the person's rejection of the molecule as other.

Lonberg was helping give birth to a new class of drug called monoclonal antibody therapeutics. It is the most important class of drugs of the last twenty years, and at this rate is likely to impact most of our lives before we die. It was life-changing for Jason, Linda, and Merredith, and for many others. Sales of monoclonal antibody drugs hit $87 billion a year by 2015 and are projected to reach $246 billion a year by 2024.

As a recap, monoclonal antibodies are exact copies of antibodies. Antibodies are essential pieces of the immune system. They sniff out and bind to antigens on other cells, including bad actors. If you know what an antibody does and create lots of copies of it, you can theoretically create a drug that fills a human being with the correct antibody and then prompts a targeted immune response.

This might sound logical after all you've read, but it still is insanely complicated, and requires high levels of both innovation and technology. So perhaps it's no wonder that Lonberg relocated to Silicon Valley, where the biotechnology business—pairing medicine and high tech—was exploding.

Lonberg's contributions wound up being significant because he helped solve the vexing challenge of how to manufacture lots of human antibodies. Lonberg's solution took years to develop, until the mid-1990s, and entailed creating what he called a "frankenmouse"—part mouse and part human. The part that was human was the immune system. Lonberg and his team could inject said frankenmouse with a particular molecule and prompt the reaction and production of antibodies. In cinematic terms, a molecule was injected into the mouse and began circulating in its Festival of Life. This would prompt the immune system to react. As part of that reaction, the mouse would generate antibodies targeted at the molecule that had been injected. In this way, the mouse was turned into a monoclonal antibody manufacturing plant, a robo-immune system, a prosthetic or synthetic elegant defense, a targeted therapy to do inside a human body what the body seemed unable to do on its own. From this, a drug could be developed built on the extracted monoclonal antibody.

But there was a twist, one essential to saving Jason. The antibody they ultimately harvested didn't target the cancer. It targeted the immune system.

For centuries, the fight against cancer had been built on the idea of attacking the cancer. But Lonberg and the company he worked for (through acquisitions, he by then was employed by Bristol-Myers Squibb) were developing an antibody that did not rely on this core idea, at least not directly. The specific antibody they were developing was aimed at attaching itself—binding—to cells in the immune systems of people like Jason.

As counterintuitive as it sounds at first, it makes a ton of sense. After all, one of the key reasons that Jason's cancer was out of control was that his immune system was standing down. It had received a signal to stop from the cancer. The drugmakers wanted to interrupt that signal in a systematic way, block it, by shielding the T cell receptor from receiving the signal to stand down.

Lonberg offers his own cinematic description of this process. He pictures a T cell, roaming the body, and it has powerful cannons on its surface. The job of this artillery is to take out dangerous organisms. But the surface of the T cell also has many antennae. The antennae receive signals from other parts of the immune system authorizing the T cell to fire or, as often as not, telling it not to fire. The cancer had succeeded in connecting to an important antenna, or maybe several, that had hit pause on the cannon.

So Lonberg and his cohorts wondered if they could use an antibody to block that antenna from getting a signal.

Their technique built on the work of others, like the discoveries of Allison and Krummel at Berkeley. Recall that these researchers had discovered that T cells could be sent into attack mode or slowed down, depending on what signal they received. The researchers had also found specific places on the T cell that were involved in receiving these communications and specific molecules responsible for sending the communications.

One way to think about the research is to picture a simplistic version of the immune system's interaction with a cancer cell.

After the cell develops, it might well have contact with a dendritic cell. This is a cell in the immune system that carries pieces of a foreign organism back to T cells for examination. The dendritic cell acts as an intermediary between a potential pathogen and a T cell. In many cases of malignancy, the dendritic cell carries back a signal that is interpreted by the T cell as a "go" or "attack" signal. The T cell then proceeds with an attack.

But some cancers, like Jason's, wind up getting a signal to the T cell that instructs it to stand down. Plus, it appears that these cancers manage to send such a powerful signal that they overwhelm the communications system; the T cell isn't really able to pick up any "go" signal.

Lonberg, among others, wondered if it was possible to displace the "stop" signal by in effect sending a louder one—swamping the T cell's "go" antennae such that it received the signal to attack. With help from the mouse, they would send molecules in to take back the T cell's antennae, prevent it from being hogged by the cancer's insidious signal, and allow it to proceed.

(For those interested in the details, Lonberg and his peers, in the late 1990s, were figuring out how to cause the T cell to receive its signal at CD28, which is the spot where the "go" signal is received, and not at CTLA-4, where the "stop" signal arrives. Both receive their signal from the molecule B7-1; if B7-1 binds to

CTLA-4, the immune system stops, and if it binds to CD28, the attack goes forward. In some cancers, "CTLA-4 is hogging B7," Lonberg said. So the goal was to "displace" B7-1 from the CTLA-4 so that CD28 could bind. They did this by creating an ultra-specific antibody to bind to CTLA-4. When the antibody bound to CTLA-4, it pried loose the B7-1. Now the brakes would have been turned off. The immune system could attack the tumor as if it were foreign and dangerous, not innocuous and self.)

If it worked, this would unleash the immune system to operate the way it was intended to work. The theory is a marvel. Within days or weeks, the body's own defenses could destroy a tumor that toxic chemotherapy couldn't kill over months or years. The locks would be taken off the T cell guns, the cannons unshackled, cancer's magic trick exposed.

The clinical tests that took place in 2007 were reported in a *New England Journal of Medicine* article published in September 2010. The drug was given to 676 patients with stage 3 or 4 metastatic melanoma, a cancer that is more or less fatal. The drug had extended the average life expectancy to 10 months, up from 6.4 months. That may not sound like much. But it's 40 percent more life!

There was a catch.

The study published in the *New England Journal of Medicine* in 2010 alluded to side effects that showed up in 10 to 15 percent of patients. Serious, serious side effects. Seven patients died, several "associated with immune-related adverse effects."

The drug, called Yervoy (ipilimumab), took the brakes off the T cell. But remember, there are lots of good reasons those brakes are there. Now, with the immune system unleashed, it could go off half-cocked and attack much more than the cancer.

It wasn't the first time that researchers had monkeyed with the immune system and paid the ultimate price.

In the spring of 2006, at a hospital in London, a handful of patients "took part in a study that has sent shock waves through the research world," noted an article in the *New York Times* by my then-colleague Elisabeth Rosenthal. The phase I clinical trial was for a monoclonal antibody that also worked on CD28. The goal of a phase I trial is to test for safety. So the volunteers at Northwick Park Hospital were all healthy, but also were selected because they had CD28 receptors similar to those of people with rheumatoid arthritis and B cell cancer.

Let me pause a moment to underscore the fact that these drugs are designed to work on two classes of diseases that, on their faces, have no relation—cancer and autoimmunity. Of course now it's clear that they are highly related. One dupes and slows or stymies the immune system, as in Jason. The other overheats the immune system, as in Linda or Merredith. Could the same kind of drug target the immune cells to restore balance?

Not in the case of TeGenero. That was the name of a drug involved in an infamous clinical trial.

Six healthy individuals entered the phase I clinical trial. They were given an infusion of the drug at a tiny dose—500 times smaller than was shown safe in animals.

"Minutes after the first infusion," one case study reads, "all patients started suffering from severe adverse reaction resulting from rapid release of cytokines by activated T cells."

Time to define a rightfully scary term: cytokine storm.

Remember cytokines? They are proteins that send signals to the immune system, creating a powerful, virtually instantaneous telecommunications network, the envy of even the fastest Internet provider or connection. The commands they send can prompt a range of responses, including cell growth and inflammation. They call on the interferons, which are central to the innate immune system, and the interleukins, which have an even broader charge, and the chemokines, which can recruit macrophages and neutrophils. A cytokine storm occurs when the network begins sending a flurry of messages, an out of control torrent of signals. The term *cytokine storm* actually understates how dangerous it is. Cytokine typhoon or hurricane might be more accurate. It is deadly.

Within eight hours, all six patients involved in the TeGenero clinical trial were in the intensive care unit.

Five of them died.

It was, to say the least, a bridge too far, a poorly constructed trial that showed how close the guardrails are when innovation gives momentum to our elegant defenses. You mess with the immune system at your own risk.

By the time Jason got sick, the drug developers had made major strides.

41

The Breakthrough

Years ago, when the *New York Times* first started putting color pictures in its paper, I joked that people should not worry: The writing in the Gray Lady, I told them, would remain dry and lifeless.

I meant it with love, of course. There's a time and place for hyperbole and exciting bold adjectives, and newspaper writing about serious subjects is not one of them. So perhaps it's understandable the way that the *Times* described with due caution what might, eventually, be seen as cancer's version of the Apollo missions. One took place on March 25, 2011. That day, the Food and Drug Administration approved for use in people with melanoma, that deadly skin cancer, the drug called Yervoy I mentioned a few pages earlier, made by Bristol-Myers.

An article ran in the *Times* in the business section, written by an encyclopedic colleague of mine, since retired, named Andrew Pollack. The story explained that Yervoy had been approved for use in metastatic melanoma, a major breakthrough. The article explained that 20 percent of people in the trial who took the drug lived two years or more. Yes, there were side effects, but not treating metastatic melanoma came with its own likely terminal side effect.

So for people dying of melanoma, Andrew's article might just as well have read: WE CAN BRING YOU BACK FROM THE DEAD!

Looking back, too, there's just no way to downplay the wording Andrew used to describe Yervoy: "a novel type of cancer drug that works by unleashing the body's own immune system to fight tumors."

This is where all the scientific study had been leading, from Metchnikoff and Ehrlich to Jacques Miller and Max Cooper, Peter Doherty, Tonegawa, and on and on. One discovery on top of another, one technique following the next, one painstaking failure leading to tiny breakthroughs, all on the backs of patients who willingly took their chances, let themselves be transplanted (begged for the treatment!) or tested with new medications, so that the immune system might not just be understood, but "unleashed."

Science and market forces had collaborated to bring a seeming miracle cure to market. Just in time for Jason.

42

Jason Races Time

After Jason's first round of chemo failed to work, his care was moved to Colorado Blood Cancer Institute, and came to be overseen by Dr. Brunvand, Jason's oncologist. The second level of treatment is called salvage. It's more toxic than the first kind of chemo. Jason responded. But there is another step in this second stage, and it is brutal.

What follows is a bone marrow transplant known as an autologous hematopoietic stem cell transplant. This transplant replaces stem cells in the patient's bone marrow that have been damaged by the chemotherapy. In a very real sense, it involves pulling out the patient's immune system and then restarting it.

That's not the horrible part. What makes this process so devastating is an interim step known as BEAM. This is another level of chemotherapy—a terrible, evil nuclear-winter-level therapy—that is used to wipe out the last of the cancer cells that are left behind by salvage. Typically, the salvage therapy leaves behind about a million such cancer cells. BEAM is toxic enough to get these remaining dogged cancer cells, but it is also so toxic that it wipes out the patient's own stem cells.

"All his stem cells are sacrificed on the altar of killing the last cancer cells," Dr. Brunvand explained.

BEAM, coupled with the emotional challenge of transplantation itself, is so intense that the procedures don't go forward until

Dr. Anthony Fauci, head of the National Institute of Allergy and Infectious Diseases at the NIH, and Dr. Mark Brunvand, Jason Greenstein's oncologist. (Courtesy of the author)

a patient is assessed on three levels: Is he responding to the chemo, and is he both physically and emotionally able to survive the experience?

It was time for Jason to be assessed by a psychologist.

On November 16, 2011, Jason walked into a small square outpatient meeting room, eight feet on each side, with a round table in the middle. He was greeted by Andrea Maikovich-Fong, a psychologist specializing in counseling people with cancer. When Jason walked in, he didn't look or act like a guy who was sick. He wore sunglasses, and not long after meeting Maikovich-Fong, he started playing air guitar and singing a rock song.

"He was very much alive," she recalled of the moment. "I don't think I'd have known he had cancer."

He was ready to take on the beast.

To prep him for transplant, Jason was given drugs that spurred growth of his stem cells and caused them to leave the bone marrow and flow into the bloodstream. That way, they could be harvested. Then it was time for BEAM.

Jason began the high-dose chemotherapy to eliminate the last of the cancer and immune function on November 21, 2011. Eight days later, after a "day of rest," he was infused with new stem cells.

At this moment, his body's immune system was laid waste, as were virtually all of his rapidly dividing cells: His gut had gaping holes, his skin could not heal, his thick hair was falling out in clumps, and his smile was nonexistent. He'd lost the optimism.

"He wore a zip-up hoodie. He was sitting there, completely hunched over," said Maikovich-Fong. "The room was all dark, and he looked like this shadow sitting there. He looked up with his eyes and not with his chin, and he said, 'This is terrible.'

"He was a completely different person than when I first met him. It was a pretty striking image if you wandered into that room."

He got out of the hospital a month later. By January 2012, Jason had learned that the BEAM and transplant appeared to have worked. Jason now was in possession of the immune system of a veritable newborn. Dr. Brunvand describes the patient with new stem cells as "like when your kids were in elementary school and brought home every virus." Jason's immune system needed time to relearn. He received antiviral medications so, as Dr. Brunvand put it, "a cold sore wouldn't turn into pneumonia," and they beefed up his microbiome with a yogurt diet—"we try to protect them and regrow the good bacteria in their gut."

Typically, the plan was to let the patient heal and then re-immunize him, as a pediatrician might a child. But Jason wasn't typical because of the fact that he'd initially relapsed so quickly, and in the same area where the cancer initially presented. "Jason had the highest relapse risk that you could imagine," Dr. Brunvand said.

So Dr. Brunvand talked to Jason about trying to cement the victory by enrolling him in a clinical trial for brentuximab vedotin. The drug is noteworthy for how it takes advantage of several of the crucial discoveries made by the immune system pioneers.

Brentuximab is a monoclonal antibody therapy, its existence drawing from the powerful discovery in the 1970s of isolating and copying individual proteins. In this case, researchers had discovered that B cells with the Hodgkin's malignancy express an antigen called CD30. Brentuximab was armed with an antibody to seek and destroy that antigen. Targeted therapy is another way to think about it.

One interesting bit of medical industry trivia arises here as well. When you see a drug with *mab* on the end, you now know it stands for monoclonal antibody.

However, just because a drug is targeted and is more precise than chemotherapy doesn't mean it is without its own side effects. Brentuximab's possible side effects include extreme fatigue, diarrhea, blood in the urine, mouth sores—the list goes on.

Jason, consulting with Beth, decided to go for it. This, he was told, would eradicate whatever Hodgkin's tried to rear its head again.

One reason for Jason's decision is that he had tremendous faith in Dr. Brunvand. Plus, Jason felt a real connection to his oncologist, someone who, like Jason, had a great sense of humor and a thirst for adventure and risk. Unafraid of a fight.

43

Shepherd of Death

June 8, 1990, was a sunny day with a lenticular cloud over the Alaskan peak of Denali, then officially known as Mount McKinley. Dr. Brunvand stood at a camp at 14,300 feet above sea level, getting ready to summit the tallest mountain in North America.

The top of Denali is 20,320 feet above sea level, with mercurial weather and unique challenges. A few months prior to this 1990 climb, temperatures had hit a record low of −57 degrees, bolstering Denali's reputation as the coldest mountain on earth. The ascent from base camp, at 18,000 feet, is actually greater than Everest's 12,000-foot vertical gain.

Just ahead of Dr. Brunvand's group of eight climbers was a Japanese team of seven climbers. They were on the mountain's West Rib, and they were in trouble. One of the Japanese climbers was suffering pulmonary and cerebral edema. Death lurked. On June 10, a message from the Park Service reached Dr. Brunvand's group asking them to help the Japanese climbers, now stuck at 19,600 feet.

Dr. Brunvand and three others ascended in whiteout conditions to intercept three disoriented Japanese climbers, reaching them just below Denali Pass at 18,000 feet. The four mountaineers forged onward and encountered two more Japanese climbers at around 19,000 feet, a thousand vertical feet from the summit.

One was a physician. The other Japanese climber was the one who had been sick with cerebral edema. He was now dead. Dr. Brunvand and two of his fellow climbers placed the body in a tarp fashioned into a toboggan and dragged it to Denali Pass, where the body was stored until a larger group could move it farther down.

In the end, Dr. Brunvand found himself attempting a rescue of someone at tremendous health risk, getting to the person too late, and winding up as death's shepherd. This sounds a lot like the job description of an oncologist.

Dr. Brunvand had grown up in Denver, his father a wide-eyed entrepreneur who had a bit of Jason in him. His father, for instance, owned a car wash in 1968; half the crew that worked there were Vietnam vets and the other half were veterans of the "summer of love." Little Mark Brunvand worked there too, sometimes running the car wash himself. This was not his calling, however. That would be medicine.

In 1985, he finished his medical residency and began a fellowship in immunology under Dr. Anthony Fauci at the NIH, eventually spending three years in Dr. Fauci's laboratory of immune regulation (small world!). Then it was off to Seattle, where Dr. Brunvand started working with cancer patients. He faced a crossroads of whether to pursue research or continue to treat patients in a clinic. This can be a tough choice for many doctors initially drawn to research, which can be seen as a particularly noble job and thus can be ego-gratifying. The top thinkers in medicine, people sometimes say, do research *and* treat patients. But that's just what people say, and it's simply false; doctors, like lawyers or writers or businesspeople, have particular tasks they are drawn to and that they do well.

When Dr. Brunvand thought about why he preferred to deal with patients, despite the intense suffering it entailed, he came down to a simple answer. "I could connect to them."

He felt that he understood what it was like to deal with difficult circumstances and loss. Dr. Brunvand had found himself— his voice in this world. It was true to him, the person who liked to connect and to feel connected, and it was an authentically heroic one. He loved the challenge of fighting on behalf of patients, but more than that, he loved "coaching" patients to deal with heinous malignancies on their own terms. On his wall at home hangs a picture and a letter written by a little girl (I've copied it here, spelling warts and all).

Dear Santa,
   I have been a verry good girl this year. I have a lot of things
I want this year
   Hear are some: 1. Poo-chi. 2. Wove love. 3. Super Soft
Kelly doll. 4. Super Poo-chi. 5. CD-Bah Hah men. 6. Tekno.
Love, Katie

Katie wrote a second note too, also on Dr. Brunvand's wall. It reads:

Dear Santa,
   Never mind all I want for Christmas is for mommy to get better and thats all.
Love, Katie

Katie's mommy did not get well.
"She died," Dr. Brunvand said.

Dr. Brunvand is both a highly cerebral oncologist and someone who owns the fact he has insecurities. He developed coping mechanisms, including humor and another trait essential to his authentic self, namely tenacity. On his high school ski team, he made himself run until he vomited or passed out, proof to himself he had trained hard. He took the cancer fight personally. "Once the decision was made to fight," he wrote, he would "try every ethical, medically sound method to win.

"If I was found to have 'cheated' against cancer, I was not going to jail but to Stockholm."

But it was the nature of the beast for an oncologist: Dr. Brunvand often came up short. He sometimes wondered if that was just his job, to fight and fight and fight, to play the martyr. The more desperate the cause, the more he dug in.

He was determined to save Jason, feeling like Jason was part brother, part son, a true fellow traveler.

44

Trials, Personal and Clinical

In May 2012, Jason added a new drug to his regimen, an antidepressant called citalopram, or Celexa. Said Dr. Brunvand: "If you have multiply recurrent Hodgkin's lymphoma and you are not depressed, you are not paying attention."

Jason had, for the moment, shaken the malignant B cells. But the process of fighting, even for a born fighter, ultimately takes its toll. Each month, it seemed, there was a growth in the list of drugs that he had to take to counteract or compensate for some other treatment. He told me he considered this regimen to be a kind of tether on his freedom. But the truth is, he probably was feeling anxious and depressed for all the reasons we know about: He sought balance even as he struggled with sleep, self-doubt, and fear, and he desperately wanted to be his old confident, athletic early-adolescent self, before the constant threat of death changed his sense of the possible.

That year, 2012, saw the science of immunotherapy continue to grow by baby steps and, at times, leaps. These developments crowned a century of learning about the immune system and were the seeds of Jason's miracle. But the progress that had been made was largely unknown to most or little regarded by all but a handful of scientists and oncologists, and probably some in the investment community.

For instance, a study began on September 26, 2012, to determine the effectiveness of Yervoy, or ipilimumab, in combination
with a new immunotherapy drug called nivolumab on patients with
advanced liver cancer. It looked at safety and effectiveness, comparing the impact of the drug on these cancer patients who had
hepatitis B and hepatitis C.

A phase II clinical trial had begun in April at MD Anderson in
Texas to explore the effectiveness and safety of the combination of
these drugs in fighting uveal melanoma, an eye cancer.

In May, Bristol-Myers Squibb launched a phase I clinical trial
aimed at studying the impact of nivolumab on patients with blood
cancers, non-Hodgkin's lymphoma and Hodgkin's lymphoma. In
phase I, the primary question was whether the drug is safe. The
trial was not scheduled for completion until 2020. By Jason's standards, that was a long way off.

Meanwhile, these were just a handful of a growing number of
trials for a growing number of immunotherapy drugs.

Some stories blow the mind, like one written later that year
by my colleague Denise Grady, an extremely insightful and deft
writer with whom Andrew Pollack and I would eventually team up
to write about immunotherapy for the *Times*. Denise's story was
about a girl named Emma Whitehead, who had been six years old
in May 2012 and suffered from late-stage leukemia, and after two
failed chemotherapies was, as Denise wrote, "out of options." She
was going to die.

Understandably, faced with death, Emma and her parents embraced
a highly experimental treatment, one that stood on the shoulders
of not just cancer research but of HIV research too. Millions of the
girl's T cells were removed from her body. Then a new gene was

inserted into the T cells. The inserted gene came from disabled HIV. Why? Because HIV is very good at attacking B cells; that's what makes it so dangerous.

But in the case of Emma, her B cells had grown malignant. That critical piece of her immune system, now a deadly force eating up her body from the inside, needed to be killed off by a portion of the immune system that remained healthy.

The new, altered T cells were injected back into the girl. They went to work. Specifically, Denise wrote, the T cells used HIV's once-deadly targeting mechanism to seek out a protein called CD19 on the surface of B cells. Think of these T cells as guided missiles programmed to find and destroy a very specific site on B cells. The catch is that the T cells weren't differentiating healthy from malignant B cells. It killed all of them.

With her B cells under massive attack, her defense system went, to use a nonclinical term, berserk.

What was happening, Denise wrote, was a cytokine storm. Denise's evocative story explained that the girl's temperature spiked to 105 and "she wound up on a ventilator, unconscious and swollen almost beyond recognition, surrounded by friends and family who had come to say goodbye."

Steroids, which as you now know are used to tamp down an immune response, failed. The doctor overseeing the pioneering effort, himself an immunotherapy legend who shares the same status as innovators like Jim Allison, had one last idea. The girl was given a drug normally used to treat rheumatoid arthritis.

"Within hours," Denise wrote, "Emma began to stabilize. She woke up a week later, on May 2, the day she turned seven; the intensive-care staff sang 'Happy Birthday.'"

The novel treatment worked. The little girl survived the side effects and joined the growing lore of immunotherapy.

But here's the thing: If you pull back the lens, this wasn't just a tale about cancer. The main character in the story was the immune system, its power to save and to destroy. Though on its surface this narrative seems to be about cancer, it actually weaves together the relationships among cancer, autoimmunity, and the most basic immune system functions, like fever and inflammation, gone awry.

In July 2012, Jason was in the middle of his brentuximab trial. He felt like he'd gone to hell. "Worse than you can imagine," he would tell me. "Nothing you ever want to go through."

Every twenty-one days, he'd return to Denver for another treatment, recover as quickly as he could force himself, and then get back to Vegas or the road, and his dreams. His casino-trinket business was doing reasonably well. The trinkets, small crystal animals like a pig or ornaments, would come stuffed with a card that could be redeemed at a particular casino for cash. The casino gave the promotions away to draw in new customers. Jason loved coming up with new trinket options, a train car for a casino in Colorado, and driving to casinos to convince management to sign up. He never could nail a Vegas casino, despite living there, and worked instead with smaller casinos in places like Mississippi and Colorado.

In 2012, he got a new business idea that came from an observation made by Beth. She was getting tons of Amazon packages on her doorstep and wondered if there were a way to keep them locked up or protected when she wasn't home to accept them. Jason became enthralled. This was it! The next great idea, a functional and aesthetic lockbox on the porch to cater to the new economy!

Chemotherapy be damned. He went looking for prototypes

everywhere—Home Depot, local hardware stores. He put a proto-type, a box with a lock on it, on his mom's doorstep in Denver. He went into remission. He was hobbled, but back.

On October 3, officials from the Food and Drug Administration met with officials from Bristol-Myers Squibb. That was the phar-maceutical giant that had, through one business maneuver and another, wound up acquiring the company and intellectual prop-erty that Nils Lonberg had pioneered. The subject of the meeting was how to put the new immunotherapy cancer drug nivolumab onto the fast track.

Fast-track designation is increasingly utilized as a way of push-ing drugs to market when there are few, if any, alternatives for patients with fatal conditions. In this case, nivolumab was in late-stage trials to deal with melanoma, skin cancer, one of the deadli-est of the malignancies when it is not caught early and surgically removed. At the time, the survival rate for the small portion caught after the cancer had spread—become metastatic—was 16 percent.

The immune system was the crux of the problem. It was para-lyzed by the cancer. This could involve two key braking systems that I've described: CTLA-4 and PD-1. The former, when acti-vated, would dampen an immune system response. The latter, pro-grammed death, would actually cause an immune cell to implode, in effect dampening the response.

Early clinical research was showing that nivolumab was help-ing turn off the brakes by turning off the programmed death response. It had been only seventy years since Jacques Miller dis-covered that the thymus, far from being vestigial, was the epicen-ter of T cell development, and now scientists were tinkering with the T cell on a molecular level. With significant success. A trial that began on December 21, 2012, and ran through most of 2013 involved 631 melanoma patients in fourteen countries, and found a response rate of 32 percent.

The FDA's decision was not clear-cut, though. It had to consider a core question of side effects that happen when the immune system's brakes are dulled—rash; cough; lung infection; colon, liver, and kidney damage; and cerebral edema, which is brain swelling. "The toxicity profile of nivolumab includes serious risks of autoimmune-mediated organ toxicity, which can be fatal and requires treatment with high-dose corticosteroids," the FDA wrote in a paper summing up some of the issues.

As we've seen, taking off the brake can send the immune system roaring ahead, which can be dampened by steroids and in turn can so dampen the immune system that it becomes susceptible to infection.

Once more: Tinker with the immune system at your own risk.

But it sure beats dying. Besides, this was still very early on. There was much work to be done.

None of this was anywhere on Jason's radar, or for that matter, on that of most people. Immunotherapy was being followed mostly by investors. They could see the potential for a series of drugs that for now were focused on a few cancers but eventually might be much more far-reaching, including that 10 percent of Hodgkin's patients like Jason who had fallen through the cracks of traditional chemotherapy and radiation.

Jason and immunotherapy were headed for a date.

45

The Other Shoe

On December 11, 2013, Jason came to Dr. Brunvand's boxy office at the Colorado Blood Cancer Institute for an eighty-minute meeting that seemed slated for joy. Jason had survived more than twenty-two months without a relapse after taking brentuximab during 2012 and experiencing a full year of remission in 2013; at twenty-four months, he'd have hit a state of remission considered significant and predictive of full recovery.

"How are you feeling, Jason?"

"All right. Good days and bad days. Some days are awesome. I get a ton of shit done, and then I'll just be exhausted."

Understandable, Dr. Brunvand told him. His body had been through three years of hell. But now Jason was down to taking only acyclovir, a medication to prevent cold sores and worse.

"You're turning the corner, Jason."

He just had to make it another six weeks and the ordeal would be over.

A week later, back in Vegas, Jason got a massage. He woke up the next day with swelling under his left armpit, where the armpit meets the shoulder. It's known as the axilla. It lasted a month. He came back to Denver to get it checked out. Now he was only a few weeks from being in the clear. He got a scan of the inflamed area.

On February 2, Jason was pumped up. He was feeling good and going to make a night of it with his oldest friend (and part of the high school gang), Bob Nesbit, and watch our beloved Denver Broncos play the Seattle Seahawks in the Super Bowl. The bad news: The Broncos got destroyed, 43–8. The good news: It was over pretty quickly, practically by the end of the first quarter. Jason had a blast with Bob and was feeling pretty good, despite the outcome of the game.

The next day, he was at the grocery store in Boulder with his mother when his cell phone rang. It was Poppy Beethe, Jason's nurse from the cancer center.

"Hey, Poppy! What's the word?"

"Jason, I've got some bad news."

"What? Tell me."

"The tests have come back. Why don't you come in and meet with us."

February 11, 4:30 P.M., Jason went to Dr. Brunvand's office to learn his fate. "I've got it, don't I?"

"Look, Jason, we're not out of options."

They were long past pussyfooting around.

"We have options," Dr. Brunvand repeated, "but there is no standard therapy at this point."

"So what does that mean?"

"You made it a year disease-free after taking brentuximab. So we can try that again."

Jason listened and just felt so beyond defeated. He couldn't begin to stomach the idea of putting himself through the hell of chemotherapy again when the cancer kept coming back.

Dr. Brunvand explained that there were two other drugs that might help. One of them was aimed at causing the cancer to over-express a molecule called CD30 so that it could be more easily targeted by the brentuximab.

"They have risks, Jason. They can actually lead to other malignancies. But those risks aren't any worse than the malignancies that you've already got."

"And if I don't do it? If I can't put myself through it again?"

"Median survival is less than six months."

Dr. Brunvand proposed they start in six days. He had the impression that Jason was on board.

"Rick, I can't do it."

He called me and told me he was done. Enough already. The guy I had known as the ultimate competitor was ready to redefine the battle. It wasn't about fighting anymore, it was about not suffering. It was about peace.

"I won't spend the last months of my life feeling like shit."

"I feel you, Greenie. It makes a lot of sense."

"It's just so wrong. I was so close. I feel fine. I feel great."

He said he wanted to talk about it some more.

Two of Jason's best friends, Noel and Tom, and I started a text conversation. We decided that it was time to get the gang together, the members of the Concerned Fellows League. We set a date. Tom would come from Minnesota and I from San Francisco, and Noel would host in Boulder with just a few of Jason's oldest and closest. We told Jason that we were having a reunion, without putting so fine a point on it: We'd be coming together to say goodbye to Jason.

Bob Nesbit picked me up at the airport and we got to Noel's house in Boulder by early evening. Tom was there, and Ariel Solomon, an absolute prince of a human being who was a year behind us in high school but had long since been part of the gang. Ariel had been a lineman for the Pittsburgh Steelers, had a Super Bowl ring to show for it, and had become a triathlete and still looked the part of a fit giant. In fact, everyone looked and seemed great, just older versions of themselves.

There were some differences. The heavy drinking that at times accompanied our adolescence was mostly gone. Two among our group had dealt with drinking problems, and they went heavy on the seltzer water. These were now open secrets and what dawned fairly early in the evening was some good news. We all seemed much more at ease with our more mature selves than the alcohol-fueled versions that we'd passed off in high school and college. We had a good conversation about family and life, and we waited for the guest of honor.

And waited.

At some point, Jason called or texted to say he was getting there. He'd driven since dawn from Vegas.

He got there by nine, a whirl of Jason, flip-flops, jeans and a flannel shirt, stale smell and a wide smile and a high-pitched laugh, and he immediately launched into a story.

"You guys gotta hear this," he said. "I was at grief counseling last night with Beth, and I was trying to use it to break up with her."

"C'mon."

"I'm serious. I keep telling her that she doesn't need this shit. But she won't have any of it. I was thinking maybe I could get the counselor on my side."

"You're a savant, Greenstein. No one else could use death counseling to try to get out of a relationship."

"I know. But it's not working."

It was beautiful. Not Jason's tactics—just Jason being Jason—smiling, laughing (squealing), not taking himself too seriously, late to his own party, living on his terms. He looked damn good.

He had another story to tell. He'd gotten himself locked up by the Vegas cops. What had happened is that over the years he'd amassed a bunch of parking tickets that he'd failed to pay because—well, why bother? Then late one night, with insomnia, he'd gone for a walk. It was hot and he was sweating like hell. He'd wound up in a so-so part of town and was walking sort of aimlessly when he realized that a cop had taken notice of him. Jason was dripping sweat. One thing led to another and the cop wound up checking Jason's info and discovering he was wildly in arrears on parking tickets. The cop took Jason to the clink, where, exhausted and overcome with the cancer sweats, he sat in a holding cell, waiting to be processed and taking in the tight quarters filled with macho twentysomethings on the brink of coming to blows.

"I had to take a shit the whole time, but there was this little silver bowl in the middle of everything!" Jason being Jason, he was howling, appreciating his own self-made misfortune, and we weren't sure if we were laughing at or with him. But speaking for myself, I thought he looked as alive as anyone in the room—certainly not like a guy who had six months or less.

Bit by bit, the conversation turned serious. Jason brought everyone up to speed and he recounted what he'd told a few of us: He couldn't do any more chemo.

"I want to enjoy one more Final Four," he said. "What do you guys think I should do?"

His question was rhetorical in the sense that I'm not sure our counsel mattered, or even that it should. It was Ariel who took the first crack at answering. "If it was me, I'd do everything I could to fight," he said. "If you've got a chance, you've got to take it."

Ariel hadn't been so much part of the saga the prior few years, and certainly had not heard Jason's recent misgivings about treatment, so his position was totally understandable. Others said they appreciated Jason's position. It wasn't a long conversation. Jason seemed burdened by Ariel's comments. He didn't like to think of himself as a quitter.

We stayed up late, playing pool and agreeing to get together for breakfast. We'd rekindled our group, our connection ripened by time, and we weren't quite ready to say goodbye to our founder.

At a restaurant painfully named Eggscetera, we brunched. Beforehand, Bob and Noel and I had talked about taking a moment to let Jason know that we agreed with him that it made sense to hold off on treatment. It's not that it actually made sense to us; that was irrelevant. Rather, it made sense to us that it made sense to him.

He nodded and took in our counsel. "Ariel got to me a little," he said. But he was still leaning away from treatment.

In the parking lot, he wanted a parting gift. He wanted to take a picture next to me, side by side, in profile, to see, as he put it: Which of us has the larger nose?

It looked to me like a tie—for last.

We hugged and I headed to the airport. I doubted I'd ever see Jason again.

Part VI

HOMECOMING

46

Bob

Bob Hoff, one of the oldest living asymptomatic HIV sufferers, had fallen deeply in love with one of the longer-living sufferers of HIV who was, by contrast, symptomatic. His name was Brian Baker, the disc jockey and record-store employee mentioned earlier. He had been diagnosed in 1992 and narrowly survived the pandemic thanks to the discovery of the cocktail. By this time, in 2014, they were living together and talking about getting married.

It had been love at first sight, at least for Bob.

That first sighting had been in 2001 at the gay pride parade in D.C., and Bob had seen Brian walking down the street and thought, "Look at that f-ing gorgeous man!" Bob took a picture of him. That seemed to be the end of that. The next year, Bob was in Chicago at the International Mr. Leather contest, and he saw Brian again. Bob's friends told him to stop being so shy and go up and say hello.

Bob approached, told Brian he'd taken his picture, and then explained to Brian that he liked to paint portraits. "Last year, at gay pride, I took a picture of you. Would you mind if I painted that picture?"

"It was the cheesiest pickup line I ever heard," Brian said. He fell for it.

They had been together ever since. In 2010, Bob proposed to

Brian, and they agreed to marry at some point when it was practical. On November 23, 2015, they were wed at the courthouse in D.C.

Shortly thereafter, Bob went in to the NIH for his usual assessment. He was in the waiting room when Dr. Migueles walked by. Bob jumped and shouted a greeting. He held up his wedding ring. "I finally convinced Brian the timing was right to tie the knot!"

The two men hugged. Then Bob told his wedding story, tears in his eyes.

"I was so, so happy for him," Dr. Migueles said.

Bob Hoff joyfully settled down and is still alive.

Bit by bit, test by test, Dr. Migueles and the team working at the NIH had spent some twenty years in a painstaking process of identifying the lifesaving quirk in the immune systems of Bob and other controllers. In fact, when Dr. Migueles was first hired, he'd made that inventory of possible mechanisms—strain of virus, number of T cells, genetic profile, etc.—and they'd assiduously gone down the list, eliminating those factors that weren't sufficiently distinct to explain the issue.

One clue that seemed crucial had to do with the strong association of elite control with the HLA-B57 gene, which is present in 10 percent of the population in North America, but present in about 70 percent of elite controllers. There are numerous HLA genes and genetic variants, so the enrichment of one in particular in this group is striking. HLAs, or human leukocyte antigens, are encoded by HLA genes and are central to the way the body's surveillance network distinguishes self from alien. HLA, it turns out, has to do with how molecules in the immune system present HIV to the CD8 T cells; those are the soldiers, the fighters, the killers.

HLA-B57, compared with other HLAs, might be more likely to present the virus in such a way as to provoke a more effective and lifesaving response. But B57 was not the definitive answer, since up to 30 percent of elite controllers do not have B57 and 10 percent of patients with typical HIV disease carry it.

"Genetics are operative but not sufficient," Dr. Migueles said they realized. His endgame was much more ambitious: He and the team at the NIH, and other scientists around the world, wanted to create a vaccine for HIV. To do that, they had to know how HLA-B57 was involved in fighting off HIV. Otherwise, they couldn't reproduce the results. If they could understand the mechanism, "you don't have to have B57," he said.

One by one, they ticked off the list of mechanisms they had hypothesized twenty years earlier might explain viral control *and* be reproducible in some way in the creation of a vaccine.

To understand what Bob is teaching us, Dr. Migueles contrasted him with what we now know about how most people react to HIV. Like Bob, they also recognize and confront the virus. They might even recognize and mount a response to the same pieces of virus. The key difference between the immune responses in patients like Bob and the mainstream way of attacking HIV appears to involve the quality and strength of the response. Bob's CD8 T cells proliferate, or reproduce themselves, to high levels when they reencounter HIV. As they do so, they increase their killing machinery and load their guns to become even better killers. These serial killers efficiently destroy any infected cells in their midst in a focused fashion. The CD8 T cells from most other individuals with HIV mount a much weaker response and have lower killing ability. HLA-B57 and a few other "protective" HLAs probably predispose a person's immune system to have this impressive response to the virus in a way that we still do not understand, but one does not need HLA-B57 to develop this major immune system offensive.

So at that point, the immune system makes a calculation that we now know to be central to the very essence of our defense network. It decides whether such a powerful offensive would be worth it. Would it make sense to create an all-out offensive that might destroy HIV but at the risk of creating so much damage to self that it would not be worth it? Should the immune system go nuclear?

No, it should not. At least that's the calculation, Dr. Migueles explained. The immune system decides that the consequences of such a nuclear war would be "radioactive" fallout—inflammation, autoimmunity, massive internal strife, maybe death.

So the immune system puts on the brakes.

"It gets toned down," Dr. Migueles explained. "This stunning tolerance mechanism is a way for the host to decide: 'This fight is too big. It's a fight that will kill this person.' So it settles for a less robust response. It cohabits with the virus, thinking, 'At least it'll kill me slowly.' What this research has taught me," Dr. Migueles continued, "is when you study autoimmune disease or cancer, how many similarities there are."

The immune system is making trade-offs to keep the peace, to maintain homeostasis, to let the individual live as long as is practical. It's just math.

Given what they had learned, Dr. Migueles and Dr. Connors started looking for the best approach to a vaccine. One idea was to get the CD8 cell to "ignore the inhibitory signal." Could they turn off the brakes on the immune system the very way that the cancer pioneers were trying to turn off the brakes so the system would attack cancer?

In theory, yes. But, so far at least, they can't figure out what molecule or molecular mechanism controls the braking system that is slowing the immune system in its fight against HIV.

There was another way to attack the problem, albeit a long shot.

In 2014, the NIH team helped other researchers take the lymphocytes from an elite controller and infuse them into a late-stage HIV sufferer. This notion was dangerous. That's because the immune system of the receiving patient might well reject the cells as nonself, just like any failed transplant. It was no small decision to try the experiment. On the other hand, the subject slated to get the cells had a multidrug resistant virus and few options, the kind of person who throughout history has grudgingly welcomed an immune system experiment because the alternative—likely dying anyway—wasn't so great either.

They took the cells from an elite controller (not Bob) and put them into the patient.

Dr. Migueles got a pleasant surprise. The B27-matched CD8 cells stayed active for around eight days. Plus, the study subject's concentration of HIV virus dropped twofold before returning to his baseline. "It was safe, and it appeared to have a transient immune effect on the virus," Dr. Migueles said.

It was a far cry from a cure, though, and the donor cells eventually disappeared. It has its own massive side effects, or potential ones. At the least, it continued to advance the idea that it's possible to build a mousetrap that is better than the AIDS cocktail.

Bob Hoff offers another lesson, and it has to do with the health of the entire society.

Were it not for people like Bob Hoff, the human species would have been wiped from the earth eons ago for the simple reason that the species cannot survive without diversity. After all, it was the diversity of Bob's immune system that allowed him to survive.

Imagine previous pandemics—hundreds of years ago, when there was no modern medicine. In those times, in those eras, the diversity of the human immune system gets credit for our survival. Some people *didn't* die of the Spanish flu or the Black Plague. Some people had a genetic predisposition, combined with a set of circumstances, that allowed them to survive.

From a purely scientific standpoint, Bob's legacy hasn't turned out to be the Holy Grail, not yet the way that Dr. Fauci and Dr. Migueles dreamed it might be. If Bob's white blood cells—his immune system—hold the key to a more natural way to fight HIV, the researchers haven't been able to tease it out.

But Bob's legacy is as powerful a statement as there can be about the immune system and human survival.

This is particularly profound because Bob's own diverse state—as a homosexual—left him, for most of his life, shunned, an outcast, like so many pitiful souls cursed by an ignorant society for being themselves. Now we can see, though, that Bob's diversity isn't just obviously one part of the human mosaic; it is one essential for our survival. The more diversity we have—physically, spiritually, intellectually—the better our balance. Just as in the immune system and microbiome. More diversity, more tools.

Brian Baker and his husband, Bob Hoff. (Courtesy of Robert Hoff)

Bob makes that point quite powerfully because he was shunned. "The irony or the sheer paradox is so powerful," said Dr. Migueles. "How unique his immune system is, so beneficial to humanity, and has been, yet he acquired this disease as part of a social subculture who is unjustly received, shunned."

Diversity in this context has two meanings—one physiological and one cultural—and both play essential roles in survival.

From a physiological standpoint, the broader the genetic pool, the better chance of having someone like Bob who will survive a pandemic and save a species. It's also a way to have a broader microbiome, and all the benefits of that. If you doubt this, just ask yourself why we don't allow incest. Such behavior leads to a narrowed gene pool, and survival rates plummet.

But it's also true that we need a diversity of viewpoints, of ideas. For proof, look no further than the lifesaving medicines I've written about in this book. They came from scientists drawn from the world over, bringing different perspectives and theories. Without them and many others, there well might not have been a more than doubling of the human life-span in recent centuries. We have diversity to thank.

Xenophobia, blind nationalism and racism, is an autoimmune disorder. A culture, tone-deaf in its own defense, attacks so aggressively that it puts itself at serious risk. Biology's lessons, honed like water-polished stone, teach us that cooperation with our species' diversity is undeniably key to harmony and survival.

47

Linda

On Friday, January 19, 2018, Linda walked up to the first tee at the Olympic Club, an elite golf course on the southern edge of San Francisco. The weather forecast had called for a predictable winter chill, but the sun shone and Linda felt warm in wool capri pants and a black turtleneck. She pulled out her driver.

She'd made a New Year's vow. This was to be the year she was just going to have fun on the golf course.

It has been thirty-six and a half years since she won the Ulster Open. She carried that grace and elegance again, after a horrific interlude of crippling joint disease. Outwardly, there was no sign of the rheumatoid arthritis, but her hands did show the cruel angularity of osteoarthritis, which is a different degenerative joint condition, not autoimmune, due to wear and tear. The swelling and crooked turn at the end of both her right middle and index fingers were largely coincidental with the rheumatoid arthritis.

Linda is right-handed, and the fact that the damage impacted her right hand was perversely good fortune for golf. A right-handed golfer grips the club most tightly with the left hand, wrapping the right hand over the top. Linda pulled out her driver. The wet ground was a mixed blessing; the ball would stick easier on the putting green, but it wouldn't roll as far on a drive, making the green tougher to reach in the first place.

Linda Segre, back on the tee. (Courtesy of Linda Segre)

Linda had been playing more since she'd retired in March of 2016. It had been a long journey to get to that place.

She and her husband had gotten a divorce many years earlier, her marriage suffering from all sorts of bumps and bruises, including her illness, the pace of their lives, the suicide of her ex-husband's mother.

In 2009, Linda took the position of executive vice president and chief strategist at Diamond Foods. She thought it would be less stressful than consulting. She was wrong. In 2011, for instance, when Diamond Foods announced plans to acquire the potato chip brand Pringles, she and the executive team, in an effort to vet the deal, traveled the world in nine days, stopping at Pringles plants and making related stops in Tennessee, Brussels, Geneva, Singapore, and Malaysia.

She didn't think much about whether the lifestyle would throw her back into disarray. "I was feeling good again. I thought: I've got this under control."

She was, in a way, playing with fire. But Linda wanted to make her own way in the world. She'd worked hard in her life and had financial ambitions. She wanted to "hit her number" so that she could retire with comfort and without fear. Like many women who become single—and despite having a boyfriend—she wanted to make enough money so that she would be totally financially secure.

Her staying power seemed to be the very embodiment of the dream of scientists and drugmakers who had come up with targeted drugs like Enbrel to slow the immune system. "She is a remarkable case," said her rheumatologist, Dr. Lambert.

Dr. Lambert made that comment during the very week Linda teed off at the Olympic Club, when they had met for her annual checkup. It was their nineteenth such annual review. The fact that Linda came in only once a year to see Dr. Lambert was itself extraordinary. Many people with rheumatoid arthritis see their doctors more often to deal with regular pain and debilitating symptoms.

At the visit, Linda got her lab results back. They were interesting only in how unremarkable they were. Nothing strange stood out. Doctor and patient went over the three drugs that Linda now takes: Enbrel, plus a second anti-inflammatory, and finally, a drug to keep the other two from upsetting her stomach. Linda asked for a refill of her Ambien to help her sleep when she travels. The sum is a quarter of the drugs that Linda took at the height of her symptoms.

Dr. Lambert marveled at her patient. "This is the vision I had for Linda," she said. Dr. Lambert recalled when Linda had first come in, thirty-six years old, in a wheelchair, because she couldn't walk. "She needed a miracle."

Dr. Lambert explained that Linda was one of the first five of her patients who got Enbrel. "She's the only one that's left." All four others have had to come off the drug because its effectiveness had ceased. This was news to Linda, who hadn't heard before about the drug's potential waning effectiveness.

Dr. Lambert explained that there are two theories about why the wonder drug can stop working. Either the immune system finds a way around the drug, or it develops antibodies that attack the drug.

The immune system, like an invisible pathogen, evolves too.

During the doctor's visit, Linda ticked off the complaints she has now, and they are modest. The twisted fingers from the osteo-arthritis. A touch of wrist pain from the rheumatoid arthritis. Sometimes, she said, her big toes have painful episodes. These were the toes that started it all.

"I'll be walking fine, and all of a sudden the joint will lock up and the pain will be excruciating."

"How long does it last?" Dr. Lambert asked.

"Ten minutes, and then it just suddenly goes away."

"Not hours?"

"No."

Dr. Lambert didn't think this was a problem in the grand scheme of things.

Linda asked if she was doing so well she might even be able to go off the Enbrel.

"We don't know whether you're in complete remission or not," said Dr. Lambert. She noted that the American College of Rheumatology recommends staying on treatment.

The mere question showed how far Linda had come.

"Her biggest problem," Dr. Lambert jokes, "is that she complains about her handicap."

Linda laced her opening drive that January morning. The wet ground kept it to 210 yards, plenty respectable, and straight. She pulled out a hybrid four-iron for her next shot to stop short of the sand traps that protect the green. But she didn't hit the club squarely and still needed a seven-iron to get to the green.

Twenty-two years ago, Linda couldn't have dreamed of hitting a golf club, let alone even walking to her ball or into the doctor's office, her body was so gripped by the suicide mission of her immune system. Now she walked calmly to the ball, as she'd done on her approach in 1982 in Ulster.

Gracefully, she swung back and through the ball. The Callaway Chrome Soft ball sprang into the air, destined for the green. It fell two feet from the cup and stuck. She nailed the putt.

"A birdie," she said. "Not a bad way to start."

48

Jan and Ron

How far we've come! When Jacques Miller began his quest to understand the thymus, the leading causes of death were pneumonia and flu, followed by tuberculosis. Much farther down the list were heart disease and cancer. Thanks to science, we chipped away at these diseases that had decimated generations and made of them low-hanging fruit.

The key had been to understand and bolster the immune system, do it with antibiotics, vaccines, and other medicines and surgical procedures that the system itself could not accomplish entirely on its own.

But like death and taxes, some things just can't be put off indefinitely. One is wasting brain.

As science helped us evade the imminence of the deadliest threats of the past, a potent new danger rose on the list of mortal causes: neurodegeneration. Alzheimer's, Parkinson's, sometimes Lou Gehrig's disease, in which the motor function of the brain disintegrates.

Worldwide, 47 million people had Alzheimer's in 2017, according to the Alzheimer's Association, a figure expected to grow to 74 million by 2030. In the United States, there are more than 5 million, which means a disproportionate share of sufferers are here, nearly double the rate here as around the world. That's likely because we have longer life-spans. Life expectancy had grown to nearly seventy-nine years, up from around seventy-five in the late 1980s (the opioid crisis has had a powerful negative impact here, and obesity was worsening it too). Alzheimer's was the sixth leading cause of death in the United States. It's part of the immune story.

This is what happens when you live longer. Eventually, your brain fails, even as your body goes on.

It's terrifying up close, as a growing number of people experience. Our intimate window comes not from Jason or Bob or Linda and Merredith, but from two of the scientists I introduced earlier, Jan Kiecolt-Glaser and Ron Glaser. They were the scholars at Ohio State University who had spent their lives studying the relationship between health and stress. Their own relationship with stress turned highly personal in June of 2011.

In the months just prior, Ron had become increasingly nervous when giving lectures. He made sure every idea that he wanted to speak about was on one of his PowerPoint slides. That way he wouldn't forget what he was talking about.

For years, Jan said, "he thought his memory was getting worse."

Ron was born in 1939. He was seventy-two. He looked physically fit at five-eleven, with silver hair and a touch of a belly. Around him, his colleagues and friends didn't notice anything wrong. His mother had had Alzheimer's, so he knew he might have a genetic predisposition to it.

He and Jan secured an appointment with a neurologist.

"When he wanted to make the appointment," Jan said, "I was scared to death."

Jan and Ron sat across from the neurologist, who was the head of the memory disorders clinic at Ohio State. Prior to the visit, Ron had completed a handful of tests, including a drawing test in which he had been shown pictures he was supposed to copy. It should've been simple, particularly since Ron had been an art minor in college.

When the couple met with the neurologist, they were shown Ron's drawings. "It was incredibly bad," Jan said. One of the images was an attempt at a three-dimensional box. "It was quite clear he could not copy the picture."

The neurologist told the couple "all the things they had ruled out," like a brain tumor. "You're doing well overall," the neurologist told them, "but there are some issues." He gave them a diagnosis of mild cognitive impairment.

But Jan could read upside down from the other side of the desk what the neurologist had written: probably early Alzheimer's.

Jan went home and read the research on mild cognitive impairment. She was the patient now, or the patient's wife, not the distant reader of her own work. She didn't like what she learned. Twelve percent of people like Ron would, on average, progress to full-blown Alzheimer's.

Ron seemed to be defying that. Through each of the next few years, he kept functioning. "Everybody saw him as the person they'd always seen him as," Jan said.

"Then, in 2014, he fell off a cliff."

He'd gone to the neurologist regularly and gotten tests of cognition that showed him dropping about 3 points per year on the scale they were using. But around 2014, he went from a score of 24 to 5 over the course of about a year. What is likely to have happened is that he had been so high functioning during his life that he was able to go through the motions of his life, in effect masking the cognitive decay.

When the mask came off, it was ugly. He couldn't reliably answer the phone or use a microwave, brush his teeth. He once put toothpaste on a comb. "It was really fast," Jan said, "and really awful."

This is an increasingly common experience. But what does it have to do with the immune system?

Up until this point, as I've used the metaphor of life's festival being protected by our immune system, I've essentially lumped together the entire human body.

In reality, when it comes to the immune system, one part of the body stands largely apart. That's the brain. It has proven more challenging to unpack than any other part of our elegant defense. One simple reason for this is that it's not easy to get a slice of it or to peer into it, certainly in real time.

The immune system and the brain, each by itself, are among the most complicated organic systems in the world, so dissecting their relationship has meant understanding each alone as well as their coordinated efforts.

There was a period when it wasn't even clear whether there was an immune system in the brain, at least like the one in the body. Part of the issue was a bottleneck known as the blood-brain barrier. This is a network of blood vessels that keeps close control over what flows between the brain and body, and that keeps many of the chemical reactions and other functions of the body from leaking into the brain. This has the profound and crucial function of keeping infection out of the brain. It's hard for molecules to get in and out. (Instead, the brain corresponds with the body through nerves that carry the electrical signals that control motor functions.)

But the immune system cells, so free-roaming in the body, are *generally* not crossing back and forth into the brain.

"The brain was thought to be immune privileged," said Dr. Ben Barres, a pioneering researcher in the field of Alzheimer's and, ultimately, in its relationship to the brain's own immune functions. "The brain has this special barrier. The immune system is not just leaking into the brain."

The brain has its own thing going on.

Brain 101: There are cells called neurons, which communicate through synapses. These connections have an almost magical power to create networks that allow the mind and body to work together. The result is a veritable neural symphony of chemical reactions executed in perfect unison. Think, for instance, about all that has to go right when someone walks or talks, let alone does a more complex task, like hitting a tennis serve or playing a piano or solving a math problem while writing the answer down with a pencil.

Graduate-level neuroscience: These neurons aren't the lion's share of the brain. A lot of the volume of the brain is consumed by a set of cells called the glia—comprising 80 percent of the brain, Dr. Barres told me. Broadly, glia are non-neuronal cells. These glia are central to the immune function of the brain. The glia come in three flavors: astrocytes, oligodendrocytes, and microglia.

As we live longer, these cells are going to be crucial in how we understand dementia and how we deal with it. What follows is a primer on the cells, their roles in the brain's immune function and its relationship to aging.

Astrocytes look like big stars. They play a critical role in helping the synapses communicate by enveloping them—a single astrocyte can envelop millions. "Astrocytes are orchestrators," I was told by a Stanford researcher, Dr. Vivianne Tawfik. Coordinators and organizers, packagers and bundlers. Crucially, the astrocytes also encase blood vessels, influencing blood flow. This helps dictate where and how much blood concentrates in the brain, with more active regions needing extra blood at any given moment, just like an active muscle receives added blood flow.

Oligodendrocytes help the neurons conduct faster signals. I think of them as speed amplifiers for your brain's internal communications network, like a Wi-Fi booster that brings the signal farther faster.

Then there are microglia. "They are the immune cells of the central nervous system," Dr. Tawfik explained.

Like the body's immune system, which originated to no small extent in the thymus, the brain's immune system also has origins in an organ long thought to be vestigial.

When a child is conceived, one of the first organs to form is the yolk sac. It eventually becomes round and grows to an average of 6 millimeters. It is a sort of food filter, with nutrition coming from the mother through the yolk sac and into the tiny forming life.

But the yolk sac performs another essential function. It is there, scientists discovered, that precursors to microglia originate, and from there move on to populate the brain. Once in the developing brain, the microglia play a key role. As the brain develops and neurons mature and die, the microglia consume the refuse. Does this sound familiar? It should. It is like the work of monocytes. It is phagocytosis. The microglia are eating neurons that must be pruned, and possibly synapses too.

Scientists' understanding of the role of the microglia and the astrocytes was embryonic in the mid-1990s, when Dr. Barres committed himself to trying to comprehend how this poorly understood system might be involved with neurodegeneration. Was our brain's defense somehow causing Alzheimer's?

Ben Barres was born on September 13, 1954. His given name was Barbara. He was born a girl, and from the earliest days, that felt wrong. "I realized from the time I was a few years old I was feeling like a boy," he said. Not much to be done about that. Not then. There wasn't even a language for it, so Barbara Barres stuffed down the feelings, wavered with suicidal ideations—"the typical transgender stuff you hear about"—and threw herself into a medical and science career turning stratospheric. She went from MIT to Dartmouth to Harvard and then Stanford. She became an expert on the brain.

In the mid-1990s, she read an article in the *San Francisco Chronicle* about a female-to-male activist in the region. She started to feel like she wasn't alone, like there might be an answer.

Then Barbara found a lump in her left breast. It was cancer. Barbara needed a mastectomy. She visited a surgeon at Stanford, and when he explained what he was going to do, she said to him: "While you're taking the left one off, take the right one off too."

Dr. Barres became Ben, blessed with a terrific sense of humor, and these kinds of comments prompted him to burst out laughing. The surgeon, Dr. Barres said, "was the first one I ever told."

The surgeon told his then-female patient that there was no health reason to take off her right breast.

"There's no way you're going to put those things back on me!" Dr. Barres told me he said, laughing. He'd go on to transition to becoming a man, and much later, he'd become something of an icon for the movement, even appearing on *Charlie Rose* to discuss it. With a new lease on life, having beat cancer and made peace with his identity, he committed himself to understanding the brain's immune system. He became a world authority.

Over time, Dr. Barres defined an immune system network inside the brain that was analogous to, but largely distinct from, the one functioning inside the rest of the body. Just like the body's defenses, the brain's defenses can cause issues. One particular paper explored the relationship by focusing on how mice develop glaucoma, a condition that in aging humans causes eye pain and can lead to blindness.

The mouse paper focused on a molecule called C1q that is involved in the immune system in the brain. Inside the brain, C1q binds to things that do not appear to be self. If C1q binds to a foreign organism, it can lead to an immune response and the destruction of the foreign presence.

In the case of mice with glaucoma, Dr. Barres and his coauthors found an extraordinary relationship between these immune functions and the disease. When mice develop glaucoma, it triggers the microglia to begin eating synapses, including healthy synapses. It's like the brain's immune system is turning on itself.

I asked Dr. Barres the obvious question: *Why?*

"If I knew that," he said, laughing, "I'd get a Nobel Prize."

Over time, though, he developed several well-founded theories that do seek to answer the question of why the aging brain appears so vulnerable to degeneration, not just with glaucoma but Alzheimer's and other conditions. One theory, Dr. Barres said, is that our brains wind up over the years with a lot of detritus—garbage—that needs to be eaten up. That spurs the microglia to eat synapses.

The janitor starts doing its job, but then the janitor goes nuts and starts eating everything in sight. In the Festival of Life, the janitor is not just cleaning up but also taking cups and plates out of the hands of partygoers—booting out cells while the party lights are still on.

Dr. Barres postulated to me that evolution has allowed such a process to go forward because older human beings are less valuable to the species. "There's nothing in evolution that would select for good brain health when you're aging. You're already through reproduction."

You've passed on your genes. What good, then, is a healthy brain?

This is speculation. For now, the science of the immune system and the brain is still embryonic, much less developed than our understanding of our body's elegant defense.

So for now, Alzheimer's involves lots of coping, rather than solving.

For nearly two years, since late 2015, Ron had lived in a memory care unit. He'd dropped to 140 pounds from 180. Jan came every few days, bringing him candy, usually jellies, and sat down and put her arm around him. He didn't recognize her. He usually didn't look at her. Sometimes, he talked to things that don't exist. He was hallucinating. He took antipsychotics.

"He's pretty tame and easy compared to many people," Jan said. Small comfort. "It's like most of Ron has disappeared. There's a shell that looks somewhat like him, and that's about it."

The circumstances have forced Jan to take very personally a lifetime of work about stress and health, namely her own health. "I go through spells," she told me. "I'll be okay and he'll take another step downwards and I'll get sad and depressed again."

What Jan knows from her own research, and the research she follows closely, is to keep her stress low and her social activity up, all of this having an impact on her mood and health. She keeps up a daily meditation practice, usually twenty minutes in her office. She tries to eat well, meaning greens and beans, because she believes that junk food impacts the microbiome, which in turn interacts with her stress levels.

"The gut-brain axis and the immune system, it's a very close involvement," she says, sounding as much like a patient looking for answers as she did the eminent scholar. She draws from a recent report from the National Academy of Medicine that relates physical health to strong relationships, good diet, factors you can control: "things that your grandmother or mother told you that mattered—you need to eat better, you need to move—but they are the hardest things to do when you're stressed.

"When we're stressed we don't want to reach for the greens and beans. A chocolate doughnut is really appealing. They may be briefly comforting, but they're bad for the longer-term process."

It also helps to cry, she says. She means we need to release stress. Otherwise the stress leads to inflammation, poor mood, fatigue, and heightened inflammation, and that can have lots of correlates. It can affect mood.

"Crying is beneficial," she says. It's an acknowledgement of who and where you are, an embrace of self at a given time and, as such, it helps the immune system by sparing it from having to deal with the repressed anxiety. Crying "is not fun. It's painful. But you feel better in many cases thereafter, as opposed to eating the chocolate doughnut."

These ideas are, as you'll see shortly, lessons to live by.

49

Jason Down the White Tunnel

My phone rang on March 17, 2014, at seven P.M. It was Thursday, and I was at Yancy's, a pub and sports bar in San Francisco's Inner Sunset, hanging out with my college roommate and watching the NCAA basketball tournament. My phone screen read: *Jason Greenstein*. "I'll be right back," I told Erik.

"What's up, J?"

"Who do you think is going to win the Syracuse game?"

"No clue. I'll take whoever you don't want for five bucks."

"You're on."

We wagered.

"How're you feeling?"

"Great. Sitting in a sports bar, surrounded by TVs. If I didn't know I had cancer, I wouldn't know I had it."

So it went. Every few weeks I'd hear from Jason or check in with him, and it was always the same: *I feel good and I don't want to go back to chemo and suffer.* "How are you doing, Rick?"

I would keep it brief. "All systems go."

"How is Prodigy 2?"

Prodigy 2 was the nickname that Jason had given to my son, Milo. At that point, Milo was seven, and Jason had nicknamed him Prodigy 2 because Milo was proving to be a very good athlete. Especially in baseball. I'd experienced occasions when people would pause to watch him catch and throw and hit when we were playing at the park. On his teams, he tended to play a year up and often was chosen for the glory positions—shortstop, pitcher, and catcher. Once an older girl watched him whack a ball and yelled "You're going to be in the hall of fame!" and Milo turned bright red. It was waaaay too soon to know if Milo had the gifts that Jason had, but regardless, Jason loved to see videos of Milo and hear stories. At this stage, conversing with Jason was a little bit like talking to an aging grandparent who wanted to hear about the kids but mostly wanted an ear to bend.

"What's the doc say, J?"

Dr. Brunvand had been clear with Jason that the longer he waited, the worse his chances of catching the growing tumor. But Jason was following his own gut and his internal meter. If he didn't feel bad, he wasn't going to suffer any further. He'd had enough of that.

Besides, he had the same sense of invincibility he'd always had. "I never really thought I was going to die," he would later tell me.

He surely was testing that theory.

By the time Jason decided he was ready to return to treatment, it was the leading edge of summer 2014. Jason finally dragged his exhausted, bloated, cancer-ridden body back to Dr. Brunvand. "There had been an explosion of Hodgkin's." That's how Dr. Brunvand put it in his note on their meeting that day, observing that Jason had a 10-centimeter mass on his neck, cancer of between 10 and 15 centimeters around his armpit, with the lymphoma beginning to stitch together as a quilt in and around the left side of his chest.

This time Dr. Brunvand told Jason there could be no messing around. He told Jason that the best shot was trying to use a three-drug regimen, including brentuximab, to get the cancer into remission, and then to undergo a different kind of stem-cell treatment called an allogeneic stem cell transplant. The stem cells would come from his sister. The use of his sister's cells would theoretically enable the rebooted immune system to recognize his cancer's subtle differences and attack it. His own immune system was clearly not up to the task. They'd need a stronger one, uncompromised by this virulent Hodgkin's strain, and even then, Jason couldn't get the transplant unless he was in remission first.

During this period, Jason would bring his brother-in-law, Paul, in with him for meetings and treatments. Paul was a molecular biology PhD, a patent lawyer, and a great listener with his science background. One day they found themselves sitting in the waiting room, and Jason started talking. "Do you want to know what it's like to be a cancer patient?"

"Let's hear it."

"It's like all the people in the world live in a Tahitian village on a beautiful beach. And I live in this canoe, and the canoe is attached to the pier by a rope. I can still see the village, and sometimes I'm allowed to come to rejoin the village. But I always have to go back to the canoe. One day I'm not doing so well, and I notice that the rope is longer and I'm farther from the pier. Then the doctors in the village pull on the rope and bring me back to the pier.

"Over time, I drift farther and farther from the village. After a while, I don't think I have the rope anymore. All around me are people, but they're not in canoes, they're in coffins. And I realize my canoe has turned into a coffin."

By the end of his story, he'd talked twenty minutes.

Paul was crying. "It was the most telling isolation story I've ever heard."

The oddest thing would happen, though, at these meetings with Dr. Brunvand: Paul could see Jason cling so optimistically to the rope. After Dr. Brunvand would give his assessment, Paul would think to himself: "Oh my God, that's grim."

"But Jason would always find the one to ten to twenty percent positive thing that Brunvand had said and turn it into ninety percent. I'd say, 'Jason, you're exactly right.'"

Part of Paul wanted to believe it, even did. "Cancer is a very resilient thing. It reminds me of an NCAA basketball game, and sometimes someone like Jason hits that buzzer beater."

In August, Jason finished his third cycle of chemo. The cancer, while it had shrunk in some places, had increased in others. To extend the basketball analogy, he was down by double-digit points in the game's final minutes.

On September 4, he went to Las Vegas to deal with his business, the best that he could. He came back September 20 to start more chemo.

He met with Dr. Brunvand on December 10, a year after he thought he'd at last been in the clear and free of all medications. Now he was on fifteen medicines, an alphabet soup of acyclovir, brentuximab, pain medications fentanyl and oxycodone, Zofran for nausea, and others.

The balance of Jason's immune system was more damaged than his cancer at this point. To support his defense network in its ineffective attack on cancer, he took chemo and targeted therapies that echoed throughout the body. With his ecosystem out of balance, he took drugs to dull the effects of inflammation, pain, stress, and depression, which themselves reflected the imbalanced immune system. Jason's repetitive cytotoxic chemo had damaged his ability to make normal blood cells. He was neutropenic, with a severely low level of neutrophils, the white blood cells that are the first line of immune defense. Without neutrophils, Jason could die of any infection because bacteria double every twenty minutes. Imagine that, beating cancer only to die of a common infection.

Jason's analogy built around the Tahitian village makes a lot of sense to me. Another fair analogy to what was happening inside Jason's body and to his immune system might be the war in Vietnam. Chemo was napalm, creating scorched earth. But that wasn't the real problem. The crux was that the plan to use napalm in Vietnam in the first place grew from a desperate and complex set of circumstances that seemed otherwise impossible to untangle. Of course, there was one simple answer: Stop prosecuting the war.

Jason was increasingly tempted to take that course. He was having suicidal ideations.

"When you say that, Jason, are you saying you have a plan?" Dr. Brunvand asked him.

"No. I just . . . sometimes it crosses my mind. I just don't want to die in this much pain. I can't take this."

They talked about whether to stop therapy. Jason said he wanted to go on. They'd need yet another new tactic. The brentuximab and other drugs given in combination were not working. But any new therapy could not be primarily cytotoxic, killing rapidly dividing cells, because that would further lower his white blood cells. Dr. Brunvand found one that had a slightly less toxic profile.

"I am concerned that his marrow is stressed enough and might not be able to recover from standard . . . chemotherapy."

Jason still insisted, in spite of all of it, that he would "beat this thing."

One thing he'd come to terms with was his relationship with Beth. Dr. Brunvand's doctor's note, at long last, reflected what had been true for years: "He does have a long-term monogamous partner."

He also continued to keep his sense of humor, somehow, even in his darkest, near-death moments.

On January 17, 2015, Jason, in Vegas, felt like shit. Everything hurt. Exhaustion overtook him. His vision blurred. Jason decided he probably should see a doctor. So what would you imagine Jason might do? He drove an all-nighter to the Colorado Blood Cancer Institute. He kept himself awake with chewing tobacco and a furious certitude that he could make it alive to Denver—the kind of will that once had earned him terrific respect from basketball coaches and foes. He drove over two mountain passes in excess of 12,000 feet with a quarter the amount of hemoglobin needed to carry oxygen.

Somehow he pulled his van into the parking lot of the cancer center, whereupon he passed out.

In and out of consciousness, he pulled out his cell phone and dialed the clinic. The team hustled out to the lot with a wheelchair, rushed him inside, and discovered his blood pressure and pulse were too low for the machines to get readings. They had to take his pulse manually. By the time they got to the elevators, Jason was able to get some words out. That's when Dr. Brunvand appeared.

"Hey, doc."

"Jason, tell me what's going on?"

"I've been spending all my money in Vegas on hookers." Jason smiled, let out one of his patented squeals. He was kidding, of course.

Dr. Brunvand laughed, commenting, "We have bigger fish to fry right now. You are going to the hospital once we find a bed."

"It's hard not to love a guy who sees God with one eye and the seedy side of life with the other," Dr. Brunvand told me.

Jason's red blood cell count had been so low—20 percent of normal—that he well could've died en route. The Steel Bull, as Dr. Brunvand called him, was admitted to the hospital and treated.

A few days after Jason had been admitted, he was visited by the medical team's social worker, Melissa Sommers. Jason was still in the ICU and Melissa came in, trying to be somber and compassionate, and asked him how he was doing. He started to complain. After a few sentences, he ripped off the covers and burst out laughing. "You thought I was naked under here, didn't you?"

She couldn't help but laugh.

"Sorry," he said. "I had to lighten the mood."

"I wish I could believe in God," he told me.

"Not your thing?"

"I just envy people who can believe. I can't find it. I've tried. It's not for me. I think it would be comforting. I see it comforting people. But I just don't see any evidence for it."

"I'm agnostic myself, J. No idea what's out there."

"Sometimes I think about my dad being up there, somewhere out there. I'll see something weird, like a light out on the highway, and I'll think it's a sign from him."

"What kind of sign?"

"Like maybe whether I should bet on the Broncos." High-pitched squeal.

"Love you, man," Jason said. He'd started telling his friends he loved them. This was uncharted language for a group of Colorado boys steeped in jock culture.

The end was near.

On March 4, Jason came for his regular visit. Dr. Brunvand examined him, with Poppy Beethe, Jason's longtime oncology nurse navigator, looking on. Empathy poured from her face, punctuated by eyes that mist up when she watches sappy commercials. She'd grown to cherish Jason.

He was complaining of a new symptom, pain and swelling on the left side of his chest and back.

Dr. Brunvand had a good idea what that meant, and he began feeling the weight of his own emotions. He gave Jason a broad clinical exam and discovered Jason couldn't move his inflamed left hand, with the muscles impinged by the tumor's growth into the nerve that fed them. He looked jaundiced. It was hard to hear Jason breathe on the left side, and his inhalations cackled. His skin was leatherish and discolored from his left pelvis to his left shoulder.

"Jason, will you excuse me for a moment?"

Dr. Brunvand opened the door of the boxy exam room, closed it behind him, and stood for a moment in the hallway, arms crossed. This was going to be very difficult. He took several deep breaths. He went back into the exam room and pulled up a stool beside Jason, sitting in the oversized chemo/exam chair.

"Jason, you are going to die."

Jason started to cry. Poppy started to cry.

"As your friend, it's my job to make you as comfortable as possible."

Jason knew one thing for sure. Dr. Brunvand wasn't a quitter.

Dr. Brunvand was built to be Jason's oncologist, the two of them a pair who would sweat together and sprint, climb, and fight, and not give up or in. This oncologist wouldn't say the words "You are going to die" unless Jason had reached the end.

"There's nothing left to treat you with. Chemotherapy is doing more harm than good."

Jason cried.

"Jason, do you understand what I'm saying?"

He nodded.

"I'd like to get your family here as soon as possible to talk about next steps."

"What about that one drug?"

That one drug was called nivolumab. It was leading-edge immunotherapy. It had been approved by the FDA in 2014 for treatment of late-stage melanoma. The drug unleashes the body's immune system. It's a monoclonal antibody treatment built on the shoulders of all the years of immunology, and it works by disrupting cancer's nasty trick of bringing our elegant defenses to a standstill. At the time, the drug was not approved for use in Hodgkin's lymphoma, Jason's cancer.

But also in 2014, an article in the *New England Journal of Medicine* presented powerful evidence that the nivolumab could prolong life for Hodgkin's patients. The article highlighted just twenty-three cases of increased survival in a clinical trial of patients with late-stage Hodgkin's, but the findings gave a sliver of hope where there had been no hope at all.

Jason's brother-in-law, Paul, and Dr. Brunvand had previously discussed the treatment, known as a PD-1 inhibitor. Dr. Brunvand told Jason he'd bring information about the "experimental treatment" to the family meeting, planned for the following Friday. By all rights, it would be a meeting to tell Jason's family to get ready to say goodbye.

Jason carried his wisp of himself back to his van.

In his doctor's notes, Dr. Brunvand had outlined what he'd tell the family. "The most reasonable approach at this point, as emotionally taxing as it is, is to consider Mr. Greenstein for hospice care," he wrote. "Palliative or supportive care would be another option, in which he can get transfusion support, but not be resuscitated and not receive any more chemotherapy."

In the intervening days, Dr. Brunvand planned for an "end of life" talk. He also spoke to a clinic administrator to find any loopholes that might allow Jason to take nivolumab. Merck agreed to allow Jason to take what was called a drug replacement, basically allowing a onetime exception under extraordinary circumstances. The hospital would not "mark up" the drug and the company would provide them with a free replacement dose for each subsequent dose Jason might receive.

Still, someone would have to pay for the initial dose. Plus, Jason was in such poor health that he wasn't even an ideal candidate, as Dr. Brunvand was poised to explain to the family.

The whole Greenstein clan gathered in a vanilla conference room at the clinic. The mood was bleak. Dr. Brunvand explained Jason's condition. They talked about the likely outcome. The question on everyone's minds was: How much time does he have? No one put it so finely, but the answer was that he had weeks, maybe a few months, to live.

At the meeting, Dr. Brunvand explained about the Hail Mary treatment, nivolumab. He told them that the evidence in the *New England Journal of Medicine* was not enough data for FDA approval, and informed consent would be required to begin a treatment with the drug. It was "experimental" at best, but had few toxicities compared to the deluge of prior therapy he had endured. Before Jason could receive this type of treatment, he had to understand fully the unknowns.

"Jason, you don't have enough platelets to do any treatment, and it's not approved." Platelets help the blood to clot and contribute to inflammation. To start treatment, he'd need a platelet count of 75,000, preferably, but maybe could get away with 50,000. His count was 8,000, indicating that Jason's marrow had been damaged by the years of relentless chemotherapy. If they could get the platelet count up, they could try. Cathy said of course she'd pay for a first dose. Jason didn't need much convincing, but Dr. Brunvand gave him a pep talk, reminding him of a story from Denver Bronco lore. The Broncos were in Cleveland facing the Browns in the 1987 AFC Championship Game. The Broncos needed to go 98 yards in two minutes. One of the Broncos is reported to have said in the huddle, "Boys, we have them right where we want them." The Broncos won.

Anyone for a miracle?

50

Jason Rises

It was March 13, a Friday, when Beth drove the shell of Jason for his first treatment of nivolumab.

He sat in that same chemo chair he'd sat in dozens of times. But this time the clear fluid dripping into his central line was not napalm but nivolumab, the product of decades of profound random investigation and learning about the immune system.

That night Jason attended his nephew's basketball game with a former teammate who wondered if the Steel Bull would make it through the night. He did. And the next. Beth stayed with him, his partner to the end. This was hospice, in effect, with a drug that wasn't yet approved for Hodgkin's lymphoma. It was anyone's guess. Jason made it another night, and then another.

About ten days later, Beth woke up and looked at Jason's back, where the lump had once so protruded she had lovingly called him Quasimodo.

"Jason, get up!"

"What?"

"Jason, you're not going to believe this!"

He wiped sleep from his eyes.

His tumor was disappearing.

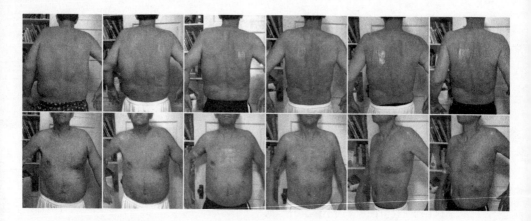

Beth, Jason's girlfriend, used her phone camera to chronicle the disappearance of Jason's tumor after he underwent immunotherapy treatment. (Beth Schwartz/ New York Times)

Dr. Brunvand's note reads: "Jason was given three doses of nivolumab"; subsequent PET and CT scans performed on April 27 "revealed a complete remission."

That's the medical speak. Here's how it sounded in human terms when Jason went for his follow-up appointment. Everyone had a different exclamation, many with expletives.

"What the f*** happened to my cancer? It went away!" he told Dr. Brunvand.

Beth asked the nurse why Jason had lost so much weight. "Because his tumor has gone," she was told. "Oh, right," Beth said, "it really was fifteen pounds."

"There was this tiny, nonscientific part of me that thought, if this crazy story is going to happen to anybody, it's going to happen to Jason," said Maikovich-Fong, his therapist. He "just has this spirit."

"In all my years," reflected Poppy Beethe, "I've never seen anything like this."

Dr. Brunvand offered up his response. "I watched the moon landing in 1969, and it was a similar sense of awe," he said. "It was with that same sense that we'd crossed a threshold. "I'd just seen the power of the immune system."

It was at this point that I picked up my pen. Could this be real? Could someone rise from the dead? Not just someone, but a close friend, someone I'd come to cherish and connect to, a person I'd watched fight and wither, and now soar into the realm of miracle. I felt like I'd seen cancer's Neil Armstrong, and a giant leap for mankind.

51

Apollo 11

If you land on the moon, you still have to get home.

52

Home

There wasn't lots of time to celebrate. Now came the nuts and bolts of hardening Jason's recovery and returning him to health.

Shortly after Jason went into remission, he got a stem cell transplant from his sister Jackie. The idea here was to give Jason a new immune system, his sister's, which in theory would be better able to fight any cancer, should it return. After all, Jason's own immune system had proven lackluster at fighting his strain of Hodgkin's lymphoma, so perhaps he'd be better off with a slightly different elegant defense.

This is a high-risk treatment. Think of it: Jason's own immune cells were removed. In their place were implanted the immune cells of someone else, a foreign presence now swarming him and playing the role of his elegant defense. His life's festival had been crashed by an alien immune system.

So it's no wonder that his subsequent medical report read: "He has had complications."

What followed was a severe bout of graft-versus-host disease. His body sought to reconcile itself to this new potentially lifesaving presence before his reaction to it would kill him.

In July, he suffered a localized relapse of cancer, about a 1-centimeter-wide lump in the skin of his right chest. It was irradiated, and again, the lymphoma responded and no other sites of disease appeared. This didn't mean that the immunotherapy had failed. Rather, doctors were now helping Jason thread an almost impossible needle. They had to suppress an immune response enough to keep graft-versus-host disease from killing him, while also keeping the immune system strong enough to fight a cancer emergence. Jason felt like he was on the gallows.

We were talking every few days. I'd gotten the idea of telling Jason's story in the *New York Times* as an intimate look into the new immunotherapy phenomenon. I broached it with Jason, and he was thrilled by this idea. It was another adventure for him, and a way to squeeze out lemonade. "Maybe it will help someone to hear my story," he said. He felt guilty that his cancer experience had so taxed the time, emotions, and resources of his mom and family, and that Beth had given so much. He allowed me unfettered access to his medical records and his doctors. Unvarnished truth. "I want to give something back."

On August 13, 2015, when I was in Denver with my family visiting the in-laws, Jason rolled up in his Windstar. He wore baggy orange shorts, a T-shirt, Ray-Ban sunglasses. My father-in-law later asked me if he had AIDS.

"Sorry I'm late," Jason said. "My mom and I had a huge fight—screaming and everything."

We sat in the backyard and Jason started to sob. "I haven't cried in a while, and the last three mornings I just sobbed. Ever since I found out the cancer is back. It's like the fifth time. No matter how many times they tell you you've got cancer, it's still a bad fucking day."

Meredith, my wife, a doctor, asked about what medications he was on, and he said that he was on these . . . he looked for the word. Finally he found it: *steroids.*

"Those can mess with your emotions," she said gently.

He rubbed his left pectoral muscle and told us how hard it is to have someone care for him, alluding to his mother. "I hate to admit it, but I still need help. Every day is hard. I HATE my life."

He talked about how frustrating it was not to be able to think about the future. One of his great joys, he explained was coming up with ideas and creating things. "But I'm not allowed to do that. I just sit on the couch all day and watch TV and take an occasional walk. What if everything was taken away from you?"

We changed the subject and reminisced about another day we had sat in the backyard—of his house in Boulder, right before the high school state championship game. Jason was ailing that day too, from a sprained ankle he had gotten when he leapt to block a shot. "I could jump," he said.

He became melancholy again. "I think how it would be easier for everyone if I was dead. But I don't want to die! I want to be alive thirty more years." And besides, this thing, this treatment, it just might take. "We'll find out in two weeks. We may have a winner."

The treatment took.

On October 5, he was back. I mean: BACK.

"Dude, I'm so fucking psyched," he told me on the phone.

His blood sugar was in the normal range, he felt good, the cancer was in remission. Jason Greenstein, a man who just a few months ago had one foot in the grave, was spinning like a top.

"I'm thinking about all these different businesses. I've had all these great ideas," he told me. "The trinket box business is really good, but it doesn't take up much time."

He began to focus on a specific idea: starting a new immuno-therapy business with one of the doctors who was a consultant with Dr. Brunvand. Maybe he'd get into the drug business.

"Dude," Jason exalted, "Dr. Brunvand had told me that I had a one in twelve million chance of being alive. I'm not trying to beat the odds. I've beaten the odds!"

At Thanksgiving, Cathy made a feast—turkey with stuffing, gravy, cranberry sauce, sweet potato soufflé, green beans, carrots and mushrooms, and pumpkin, pecan, and apple pies. Guy, Jason's brother, made extra turkey. Cathy told everyone to come early. It would be a real celebration.

"I'm so grateful. He's alive and well. It's a miracle!" Her voice rose in its high-pitched way. "I just wish Joel was here to see it."

It was just like old times, including the fighting. The Greensteins were birds of a feather, together, on Thanksgiving.

Cathy and Jason, for instance, had been bickering because he hadn't taken his cell phone to the camera store to print out the pictures of his medical journey. "That's your documentation of what happened. If you lose the phone, you've lost it."

"Let it go, Ma. I said I'd do it!"

"I just wish he'd focus on one thing," she told me, then suddenly relented. "Well, he does feel still crummy some days."

The holidays came and went and Jason had good days and bad days. The cancer was gone, but years of ingesting piles of drugs, some he was still taking to deal with side effects, had left his body depleted. In February, he came down with a mild pneumonia and needed antibiotics. He was still on blood thinners, and his nose bled. One night he was at a restaurant called the Chop House and went to the bathroom to stop the bleeding. He accidentally dropped a bloody tissue on the floor and bent down to pick it up and his back went out. The shoulder blades and upper back just seized, and he could feel it in his stomach muscles and ribs too.

Just a setback, he said. He'd already gotten a business plan written to do marketing and sales for an immunotherapy-related business with a doctor he was partnering with. "I've literally written and structured the business," he said. "I've created the brand."

Snow pummeled Denver one mid-March day. Jason went to shovel his van out of the snow so that he could go to an appointment at the clinic. But he had broken his snow shovel in a frustrated painful fit a few days earlier, so he used a folding chair to clear a path out of the garage. He arrived to the clinic wet, cold, and two hours late. His back was killing him. He was X-rayed to check for worsening pneumonia or a bone lesion to explain the back pain. Nothing showed up, so it appeared the intense pain was most likely due to shoveling the snow with the folding chair, maybe a muscle strain. At the end of his clinic visit, Dr. Brunvand drove him home. "I was certain his tires were bald and his van only had rear-wheel drive." At his house, Jason gave Dr. Brunvand a trinket box that looked like a rose. "Chicks love this box; take it home to your wife so she will forgive you for being late," he told his oncologist.

Over the next few weeks, the back pain worsened. He kept shoveling the walk at his mom's house in Denver, where he and Beth often stayed, but he continued to use the folding chair to do the shoveling. His back finally went out completely, leaving him in excruciating, immobilizing pain. He had the flu and some lingering pneumonia still. He went to the doctor and they scanned his spine. The cause of the pain wasn't clear. Dr. Brunvand suspected the culprit might be a relapse of the cancer. At the base of Jason's spine, they found what appeared to be a lesion. They weren't sure. It looked like Hodgkin's might be trying to sneak back in.

I found out when I called Dr. Brunvand on April 7 just to check in. "He's relapsed," Dr. Brunvand told me.

He favored using more nivolumab. It had saved President Carter from a melanoma relapse that had gotten into the spinal fluid. That showed the drug was capable of passing through the blood-brain barrier, so it might help Jason's spine. More nivolumab for Jason also meant revving up the immune system, and that would risk more graft-versus-host disease. The chessboard was filled with land mines on nearly every square.

"We're in uncharted territory," Dr. Brunvand told me. "I may sound like a coldhearted son of a gun, but freaking out and feeling sorry for ourselves is a luxury."

He described fighting cancer as a knife fight in which the disease keeps standing up again, attacking again. "You're hosed if you lose your kind of thoughtfulness, your desire, intensity."

Jason, he said, had to fight back. Not everyone would agree, of course. Some people fairly reject such thinking as suggesting that Jason's survival depended on toughness and tenacity, when cancer, as much as it is a knife fight, is also a game of chance. Sometimes you survive and sometimes you don't, and your commitment to winning isn't the difference between life and death.

But I took Dr. Brunvand's point. He was a fighter and, in this case, was channeling the fighter in Jason, who was back on the mat again, knife in his side.

After I got off the phone with Dr. Brunvand, something happened to me that hadn't happened during the whole ordeal. I cried.

On April 19, I landed in Denver to spend a few days with Jason. He was staying in his mom's modest single-story Denver house, beige brick on the sides, with a green roof. In the front room, Jason sat in an aging recliner covered by a towel and a sheet. The room smelled like the unfiltered cigarettes smoked by his mom. His feet were warmed by gray hospital socks. He looked like an ancient mariner, in boxer shorts, full head of hair.

"Hey, Rick." Not much life in his voice.

"Greenie. You look like shit."

"You're telling me. I think I've got a broken back."

He didn't know what was going on, and I sensed the doctors weren't sure either at that point. How could they be? Jason had dealt with so many different foes—cancer, infection, medication, graft-versus-host disease. Jason couldn't move so much as to walk to the bathroom. His mom waited on him hand and foot. The pair were in fine form, alternately gibing at and comforting each other, as in the conversation they had after she came in from having a smoke outside.

"Did you have a good smoke, Ma? It's time to take my insulin."

"This starts the minute I get up in the morning. Do this, do that."

"I have to take insulin, Ma, or I'll die."

He pulled up his T-shirt, giving way to a little paunch covered with small bruises. They were caused from the injections of blood thinner and insulin, all taken to chase the side effects from the dozens of pills he takes. "So many complications," he said. He was taking two different kinds of pills, for side effects and pain, and for side effects of the pain meds, and on and on.

"I've never been in pain like this."

"They are killing him with all these treatments." Cathy turned to me, her voice rising. It's harder for him than most, she said, because the last thing he wants to do is to be told what to do, even taking pill after pill at certain times.

"I tend to like to wing it, to be a wheeler-dealer," he said. He acknowledged he should've been better about taking his treatment regimen in the past. "I was kind of a ding-dong."

The next day, we were supposed to take Jason to the hospital to get a dose of nivolumab and have the lesion in his back evaluated to see if it was responding. As we got ready, the exchange between mother and son was priceless.

"Jason," Cathy said, "there are some things I want to ask the doctor."

Jason's face and body strained with tension. It was as if he were holding himself back on a debate stage, trying not to explode. Finally he exploded. "Ma, you're not a doctor. This is not about you."

"I know that, Jason. I'm not going to challenge them. I just want to ask some questions."

"You're not going to challenge them, Ma!"

"Fuck you, you're goddamn right I'm going to challenge them!"

Just as fast as the tension had erupted, it dissipated. "I'm going out to have a smoke."

"Good idea, Ma, have a nice smoke."

An hour later, we lifted Jason from the chair. He put his hand in mine and we walked down the stairs and I scooted him into my rental car.

At the hospital, Cathy tried to keep her cool, but it was all so damn confusing. Dr. Brunvand told them that the scans of Jason's back showed that the lesion "probably represents cancer." He thought it might explain a compression fracture in Jason's back, caused by the treatment.

It wasn't altogether impossible, though, that Jason's back had fallen apart, like brittle wood, owing to years of steroids and chemo-

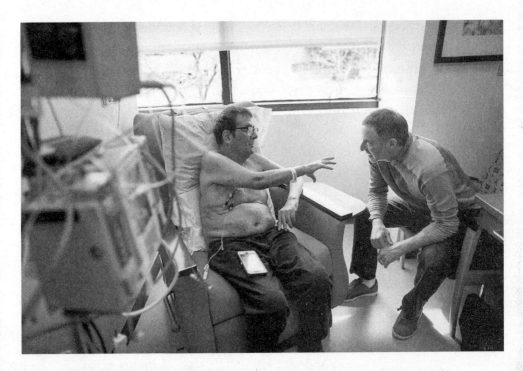

The day the author brought Jason to the hospital in the spring of 2016. It would not be the relatively routine visit they'd hoped for. (Nick Cote/New York Times)

therapy weakening the bone structure. Dr. Brunvand thought it better to be safe than sorry and continue to treat for cancer. Jason had tears in his eyes.

"You're an animal," Dr. Brunvand told him, "not a sloth, but a tiger."

Jason's pain was so bad that they admitted him to the hospital for an MRI with a high-resolution imaging of the cancer and the bones involved, as well as to administer a spinal tap to deliver chemo into the spinal column and diagnose whether cancer was growing in the fluid around the column and the brain.

The next day, Jason sounded upbeat. "There appears to be little

or no cancer in the spinal fluid. The cancer is very minimal or fad-
ing out."

Next came back surgery to repair his compression fracture.

"The news was just so great. It's incredible news," he said. "Dude,
I have another chance."

I started to have nagging doubts. Did Jason have cancer for sure, or
had his back collapsed as a side effect of something else? Dr. Brunvand told me that all signs pointed to a high likelihood of some
relapse, but it was minimal and treatable. It was the malignancy,
the oncologist felt strongly, that had caused the back to fracture.

Either way, Jason's body—the festival inside the wind-wracked
tent of his life—was way out of balance. Now, having learned so
much about the immune system, I understood how his body, kept
alive by all the medicines, was trying to compensate and overcompensate. I was having trouble seeing how Jason could find balance
again. But he was sure that one more back surgery would do it, and
he'd be on his feet again. He would fight on. The plan was to keep
him in the hospital through the surgery and rehab.

Over the next few weeks, we talked a few times and exchanged
voice mails.

5/28/16: "Hey, Matt, this is J. Sorry didn't return or answer
calls you've given. This thing is really fucked up and rough in the
hospital . . . But my spine healed really well. I really need to get
strength back to the point where I can walk and that's it and I'll
be out of here, and that's kind of where I'm at. I can't believe my
legs are so weak like they are, but I'm building strength day by day.
That's what I'm doing. I hope you're doing well."

6/1/16: "Hey, Matt, it's Greenie. I just wanted to tell you I got
some great news today. I got the results from the PET scan. It came

back totally clean! No Hodgkin's in my whole body. So it's amazing, anyway, great news. Now I gotta get out of here—hoping in two or three weeks, that's what I'm thinking."

In late June, crisis. Jason was having trouble breathing. He didn't want to eat. The nurse gave him another pill, this for a panic attack, but he still wasn't eating. They put in a feeding tube. He became unresponsive. It wasn't making a lot of sense to Dr. Brunvand, who initially noted, "His counts look perfect, and his CT scan is completely unremarkable."

Prior to becoming unresponsive, Jason had mentioned to Beth that he was giving up and wanted to die. He couldn't stand the pain and the endless hospital stays. "He has every reason to be depressed. But I'd like to get him through this. I don't see any reason he should die. I'd like not to throw in the towel just yet," Dr. Brunvand told me.

He thought Jason was showing signs of emotional distress. As more test results came back, Dr. Brunvand thought he understood the problem. The tests were showing that Jason was experiencing a spike in inflammation, a version of a cytokine storm. "A hallmark," Dr. Brunvand told me of "toxicity after nivolumab."

The inflammation, he theorized, was impairing Jason's brain function. It was a kind of coma, Dr. Brunvand told the family. They gave him steroids to slow the storm. "Let's see if this reverses and he wakes up and smiles."

He awoke three days later. He just popped awake and wanted dinner. When I got the call, I jumped up from my desk and sobbed with joy. "He's alive, Meredith. He's alive!"

He was still there in July, dealing with one complication or another. I was in Colorado and dropped in at the hospital on July 27.

Jason was weak and tired. He left a message the next day: "Hey, Rick, how you doing, it's Greenie. Listen, man, I wanted to say thank you for visiting and hope you had a good trip here. Sorry I was kind of out of it that day—it's weird—it ebbs and flows, but overall I'm great. I've got a liver biopsy today and then dialysis, so it's kind of a scary day, but I'm coming back, dude. So we'll see, man, if I can climb out. Anyway, I love you, and thanks for coming."

His biopsy came back negative for cancer but suggested he could be experiencing liver failure. After his liver biopsy Jason bled about twenty units of blood around the biopsy site; he went back into surgery to have the bleeding stopped. The threats came at every turn.

Organ failure is another sign that the immune system is attacking the body, although this was not necessarily a side effect of the cancer treatment. It could have been lots of things. Jason was told he would need to spend the rest of his life on dialysis as a best-case scenario, or he would die of organ failure before he ever left the hospital.

This news was too much for Jason. Already in desperate pain, stuck in a hospital bed, now the ultimate dreamer and pioneering soul was being told he would be a patient forevermore. That's how he saw it, at least.

"I'm done," Jason told the psychologist who delivered him his prognosis. "I gave it as good a try as anyone could."

53

Jason's Way

On August 10, the day after Jason declared himself finished, Meredith and I visited the hospital, unsure which Jason we'd meet. We saw the one who really had had enough. He was mostly unresponsive, his head back, mouth open. His mom sat at the foot of the bed and Beth sat next to him, brushing his brow.

I retold Jason some of our stories from the glory days, as if he could hear, and we all tried to laugh.

The nurse gave Jason morphine. He calmed. There was talk that he could last a few days. Cathy went to grab a quick bite.

"This is it," my wife suddenly said. Jason's breathing had become particularly labored, a pattern Meredith, a doctor, well understood.

Beth wiped the hair from his forehead and kissed him there. "Goodbye, my sweet love," she said.

Jason took a final gulp of life.

Determined as he was in life, he made up his mind, and off he went. It stood to reason he'd picked that moment so his mother, champion and stalwart, wouldn't have to bear witness.

A few minutes later, in the emotional and medical vacuum, I found myself standing alone at Jason's bed, looking at someone who had never been inert in his fifty years.

"Love you, Greenie," I told him. "I want to thank you for never looking down on the little guy. I hope my son can carry himself with that same dignity and class."

Jason's memorial service, a few days later, was powerful and sad and funny. I eulogized him and told a story about how he and Tom had once driven during college from Boulder to Berkeley in the Volkswagen Beetle Jason had inherited from his dad. Jason and Tom had blown most of their money by Wyoming, when the Bug threw a rod and they had to hire a mechanic to save the car. By Reno, they had only $50 and were low on gas. Jason decided the best thing to do was to . . . attempt to double their money in a casino. They lost most of the remaining $50 playing blackjack, slept in the car, and used their last $5 on Cool Ranch Doritos. They made it to Berkeley on fumes in time for the kickoff of a football game. I described Jason as a guy who went farther on a single tank of gas than anyone I'd ever met. I said I imagined that Jason was in that Bug up in the sky now, driving away, maybe in the direction of his dad, Joel, who was waiting in the heavens, wearing his ratty brown catcher's mitt.

I wrote an obituary for the *New York Times* to update the story I'd previously written about Jason's saga describing the potential hope for immunotherapy. After all, it had given Jason an extra year.

But what was the sum of it all, now that Jason was dead?

54

The Meanings of Life

Who do I think I am, titling a chapter "The Meanings of Life"?

This is not a typo, either. I emphasize *meanings*, plural.

I'm not so audacious as to think I can distill it all into a single meaning of life.

But I can say with a straight face that I have a decent idea of several of life's essential attributes, as seen through the lens of the immune system. This network is so central to our being, our survival, that its inner workings offer elegant lessons for living better, even living longer.

Each of these lessons comes from understanding what makes the immune system so effective. It is eons old, honed and polished by evolution, and so, by definition, very good at what it does.

First, everything is connected. Cancer, autoimmunity, HIV, the common cold, allergy. The immune system, our elegant defense, is the river that runs through every aspect of health and wellness. It tends to the festival of our lives, and it does so by seeking balance and harmony.

It seeks peace with its surrounding environment. This is a far different idea from the one I began with when I started to learn about the immune system and assumed—as I suspect many people do—that its chief jobs are to defend and attack. Defend, yes; attack, not necessarily. In fact, the immune system is constantly seeking to maintain harmony, not just by limiting its attacks to all but the most necessary ones, but also by cooperating with the organisms that surround and invade it. At its core, it tries to discern self from other, but having done so, it doesn't just destroy what is alien.

It has made allies of the bacteria that thrive within it, and the bacteria have made an ally of the host. In fact, if our immune system had gone to war with each organism it deemed as different, the species would not have survived. For us to have a fully effective immune system, we need regular engagement with bacteria in our environment and in our gut.

This realization adds a profound level of nuance to the idea of what *self* is, and what *other* is. What is *alien*, what is *foe*, what is *ally*, what is *partner*?

This teaches us clearly that our survival, as individuals and a species, is best served by cooperation. This may sound obvious, but civilization, even of late, has been dominated by the push and pull of our competing instincts to cooperate and alienate, to see what people share in common or prey on what divides them. The lesson of the immune system is that the better able we are to find common ground, the more allies and weapons we have to contend with a greater, common foe.

This is a powerful argument too for diversity. The more diverse our genetic tool kit, the more options and ideas we have to enable our common survival. Bob Hoff was the ultimate outcast, a gay man in Des Moines. He is not someone to be castigated as different, though, but to be embraced as a genetic and cultural ally, a brother, an essential part of our common survival.

The scientists from other countries created the foundation of learning that became the medicines that forestalled Jason's death, helped Linda, and that may yet come from Bob's contributions. If we learn together and cooperate, we can tackle autoimmunity and cancer and Alzheimer's, and who knows what other seemingly impossible foes.

Conflict has its inevitable place. Societies and people will collide, just as at times our immune system must play a vigorous defense. But the immune system cautions us to take the least destructive path possible to a livable balance. When we don't cooperate, when we err too easily on the side of war—literal and proverbial, physical and verbal, armed and political—we emulate one of the most self-destructive of our traits: an overheated defense system. In fact, among the biggest misconceptions that I took into this book was that it is better to have a superpowered immune system. The advertisements are everywhere urging "Boost your immunity!"

Wrong.

Dr. Fauci, one of the leading scientific lights in the world, said that when he hears ads promising to boost your immune system, "it almost makes me chuckle. First of all, it is assuming your immune system needs boosting, which it very likely doesn't. If you *do* successfully boost your immune system, you might boost it to do something bad. Even with the very dramatic positive results we're getting from immune therapy with cancers, we're looking at clinical trials with very, very toxic side effects. It doesn't just suppress the cancer but puts in a bunch of things that put system out of whack."

Some of the most devastating chronic deadly conditions in the festival that is our life arise when this system goes even a touch out of control. Fatigue, fever, stomach issues, rashes, organ failure, flooding of the lungs, and on and on. These effects are so devastating that it is difficult at some points to know the difference in the effects between pathogen and inflammation. Sometimes these effects are actual autoimmune disorders. Other times, they are episodes of overheating, the fatigue and acne and sores and leaky-gut-prompted stomach issues that kick in when our elegant defense turns into a police state.

The immune system teaches us to err on the side of cooperation and acceptance.

It is true too on the other side of the equation. If you suppress your immune system deliberately, through medication, it can mean trouble. Dr. Fauci has never treated Merredith Branscombe—the woman whose autoimmunity remains elusive—but I talked about her situation with him and he was sympathetic to her quandary. The mechanisms behind autoimmunity, for as much as we've learned, remain murky, even as the monoclonal antibody treatments have become more precise.

"Generally, you have to give broadly nonspecific suppressors of the immune system," Dr. Fauci said. "It comes with absolutely inevitable toxicities."

There's a significant lesson here for society. In our quest to build a perfect and efficient world, we have overcorrected.

As I noted earlier, it's hard to name a single profound innovation that hasn't had extraordinary side effects. When cars hit the scene, we had much greater freedom of movement and incredible new efficiencies, and also crash-related deaths soared; driving is now the single most dangerous thing most people will do today.

With the industrialization of food, we packaged and processed and transported food and got more calories to more people, helping cut down sharply on malnutrition. But our industrial processes introduced junk food, and obesity has soared around the globe, doubling in seventy-three countries since 1980, and rising in most others. Diabetes is rampant. Poor diet is killing us by the millions.

An atomic bomb ended a terrible war. The same technology leaves us in constant peril.

With television, computers, and phones, communications have become the stuff of nineteenth-century science fiction. Texts from Everest! But we are increasingly drawn to the bells and whistles, the novelty, the rush of dopamine when we stare narcissistically into a selfie camera while we are driving.

Industrial processes changed every facet of life, from clothing and housing to transportation and communications. But smokestacks led to a changing climate with apocalyptic dangers.

And there is arguably no more powerful medicine on earth than antibiotics. They are vital for our survival. Full stop. But their widespread use also threatens now to cause the evolution of bugs that will make past plagues look like the common cold.

These examples are not arguments against progress. This is not Luddite talk. But it is an argument for awareness. Sometimes we cannot control our world and hold it too tightly without squeezing some of the life from it.

In the case of the immune system, we have tried to overengineer. It has cost us. We must learn sometimes to let nature lead.

It's what Merredith teaches us. She has learned the hard way.

In December 2017, Merredith was taking a walk with her dogs, just six months after she and I had taken the walk I described at the beginning of the book, when she showed me how the sun inflamed her skin. One of the dogs, Bam-Bam, stopped suddenly. Merredith tripped over the dog and landed on a rock. Exquisite pain rocked her arm.

As she walked to her car to go to the emergency room, she could see her arm hanging so loose it was as if it were flapping in the wind.

The humerus was so shattered that it required forty-four pins and two plates. The surgeon told her that he suspected this was from all the medications she'd taken, weakening her bones.

Merredith's journey draws me back to the challenges of tinkering with the immune system. She came eventually to treat herself less with modern medicine and more so with primitive methods, the tools of our great-grandparents, herbs and rest and nutrition, vitamins, turmeric, tart cherry. These are not chosen at random, nor merely the stuff of folk wisdom. Some of these have decided scholarship backing up their anti-inflammatory properties. (She also swears by probiotics.)

She knows her triggers: sun—"especially sun"—sugar, processed foods, whey.

She has become her ship's captain. "The clues were there and I could find them. I could listen to my body. I controlled for what I could and then researched other symptoms and causes, found papers and studies showing, for example, that autoimmune patients typically have massive deficiencies in vitamin D. So I added vitamin D. I already knew anecdotally from my own experience that B vitamins could be incredibly helpful in warding off fatigue. I began adding water-soluble liquid B vitamins (i.e., MiO) to my water. And so on, all trial and error, until I cobbled together a regime that seemed to work. It's not perfect, but it's important to note that I am *not worse* than when I was on medications."

Much of what I've written here is a celebration of science and of medicines born of it. In no way do I intend these takeaways to detract from human progress. The best example is the advent of antibiotics. It helped begin a journey that leaves us now with another incredible milestone, the drug that gave Jason another year of life. I wish for everyone, for my family, for myself, the development of treatments that will prolong and enrich a quality life.

What Merredith's story illuminates, though, is that these drugs—as the immune system teaches—must also be used with an eye toward the delicate balance that has led to the survival of our species. Even now, though, we are pulling back sharply on the use of antibiotics so that the element that saves us doesn't lead to civilization-threatening pandemic.

The takeaway here is to understand the risks and the motivations of companies selling the drugs that address diseases.

"The pharmaceutical industry has made a business out of targeting them with specific drugs and antibodies. I can't stand it any longer," said Dr. Dinarello, who helped us understand fever and interleukins. "Psoriasis, arthritis, bowel disease. The industry is targeting different ways of treating them—all targeting cytokines."

But the risk is infection, even cancer. Why? Because, as you now know, you're tinkering with a very sensitive system.

"Take the patient. His immune system is pretty much under control. His physician says, 'You can feel a little bit better if you add this antibody. You do have the risk of infection, but we can take care of that,'" Dr. Dinarello says. "Patients go for the risk."

Big money is at stake, he said, and added: "Just look at the ads on TV."

The profits are huge. Decent chance that use of these could help save your life or the lives of your children or grandchildren. Better chance it'll come with side effects.

Coincidentally (or maybe not), the very night after I interviewed Dr. Dinarello on this subject, I was watching the news, and on came an ad for a drug called Otezla, which is used to treat psoriasis. The ad noted a list of potential side effects, which in and of itself is little different from many ads for drugs. Some sounded quite typical, like nausea and diarrhea, but others stood out. "Some patients reported depression and suicidal thoughts."

Now that the connections between inflammation and mood were clearer to me, these potential side effects seemed to feel more real, not just "in the head."

I went to the company's website, and that's where I found additional disclosures. The FAQ on Otezla states:

> The exact way in which Otezla works in people with psoriasis or psoriatic arthritis is not completely understood. Based on laboratory studies, what is known is that Otezla blocks the activity of an enzyme inside the body called phosphodiesterase 4 (PDE4). PDE4 is found inside the inflammatory cells in the body and is thought to affect the process of inflammation. By blocking PDE4, Otezla is thought to indirectly affect the production of inflammatory molecules, helping to reduce inflammation inside the body.

To Dr. Dinarello, the issue of the side effects of these medicines underscores a simple message: "It reveals how sensitive the immune system is to suppression."

Buyer beware. Be aware. Take care. **You tinker with the immune system at your own risk.**

For those who seek a different path, like Merredith, there are things we can manage, things that science shows us are powerful. The best examples are those over which we have complete control: sleep, exercise, meditation, and nutrition.

Sleep and exercise play such a key role in keeping the immune system in check, partly by keeping the adrenal system from firing too intensely; when it does become too intense, adrenaline—

epinephrine and norepinephrine—can create the cycle in which cytokines are released, leading to inflammation, sending the system further out of balance, even leading to more sleeplessness and more adrenaline. Not only can inflammation increase, but other parts of the immune system can become compromised, less able to function. At the very same time, the festival becomes susceptible to overzealous immune cells and to pathogens that are not held in proper check, like herpes.

The so-called type A lifestyle is a good way to let your immune system go wacky, and to no good end. Linda Segre can attest to that.

With regard to nutrition, a simple conceit: The less toxic the things you put into your body, the less likely your body is to create, or need to create, an inflammatory response. When there is an alien presence—say, cigarette smoke—it leads to a disease cascade, including inflammation and then a need to rebuild damaged tissue. The more times there is such damage, the likelier the new cells will be malignant ones, with the terrible combination that leads to successful cancer. When it comes to food, science identifies risks associated with unnatural substances you digest, additives and chemicals and factory inventions that are not actual food. They make it likelier that your immune system has little choice but to react.

There is even more evidence supporting the value of lifelong exercise. One particular study, published in 2018, shows the importance of exercise to the immune system and longevity. The study looked at the immune systems of people aged fifty-five to seventy-nine, comparing more sedentary people with regular cyclists. The people who exercised showed several crucial differences in their elegant defenses: The cyclists produced more new T cells from the thymus, and they had fewer cytokines that cause the thymus to decay. The upshot of the research is that exercise slows the natural aging process of the immune system.

These tips are well-worn, but perhaps at least you can now see the scientific basis for them and the way they connect to your immune system.

Or you can take your cue from Dr. Ephraim Engleman, who was an immunology giant who by most standards lived forever. At one hundred and four, he got his driver's license renewed. He still commuted to the office to study autoimmune disease. He died just shy of his hundred and fifth birthday. He was in his lab, at the University of California at San Francisco, where he pioneered research into causes and cures of rheumatoid arthritis. The year was 2015.

An obituary published by the university listed his self-professed secrets for longevity: *Avoid air travel, have lots of sex, keep breathing, and most appropriately, enjoy your work, whatever it is, or don't do it.*

So there's that.

I tie these points together with my own observation drawn from the sum of my research. The more active you stay, body and brain, the more you signal your internal systems that you continue to play a vital role in your own survival and the survival of the species. This leads to a virtuous cycle in which key internal mechanisms continue to regenerate, allowing you to play a vital role and, when you do that, pushing the cycle on. By contrast, if you grow stagnant, physically and mentally, the system is signaled that you are calling it quits and it need not "waste" resources on your survival.

Finally, among all these lessons is the one biggest surprise I took from writing this book. I'll call it "The Meaning of Jason."

55

The Meaning of Jason

When I started reporting this story, just as Jason had risen from his deathbed, his cancer miraculously gone, I thought I might be writing a book about the quest for immortality. The journey of the immunologists was reaching a point at which we could resurrect people. As a species, led by an international cadre of brilliant scientists, we were discovering how to tinker with the immune system, such that we could prolong life for who knew how long.

It was the first question I began asking: Are we looking at living a long, long time? How can I not wonder? Is this about immortality?

It's absolutely fair to say that the journey to extend life has been a defining characteristic of the human condition.

If the quest has been immortality, we are miserable failures. Yes, we're living longer, and better, but the best we've got to show for it is the occasional person who hits a hundred and ten. It's a blip. Now I understand one of the key reasons. Our immune system is doing us in.

You heard right. The defense network—so often upheld as the key to health, and it is that, certainly—plays a big part in the way Jason's story ended, and in how all of our stories will end.

The underlying reason for this particular meaning of life comes from several key aspects of the immune system—attributes that I've laid out over the course of this book.

One has to do with the trade-offs that the immune system is

constantly making to keep things balanced in the Festival of Life. Take, for example, wound healing. The immune system has to allow our cells to divide so that we can rebuild after injury. The immune system fosters new cell development, helps access blood and nutrients, lets the festival thrive. But this trade-off also allows for the strong possibility—even inevitability—that malignant cells will thrive.

"Cancer happens in everyone," Dr. Jacques Miller, who a lifetime ago discovered the role of the thymus, told me as we discussed the meaning of the immune system and life. The brain will fail, the organs will shut down, the lungs will flood. Some of these will owe to breakdown of our defenses, some to an overwhelming pathogen, but some, like cancer, will arise from a complicity of the immune system itself.

The reason is that the immune system hasn't evolved to defend us as individuals. It has evolved to defend our genetic material and the species as a whole. It does an extraordinary job of keeping us alive until we reproduce and then rear our offspring. After that, it does an even better job of moving us out of the way.

"Evolution has decreed we cannot live forever," Dr. Miller said. "Nature, evolution, has decreed you've got to make way for the next generation."

Ruslan Medzhitov, the Yale scholar whose pioneering work illuminated the innate immune system, echoed this thought and added a point that no medicinal fix we contrive will lead us to live forever. "There is no ultimate solution. There is no free lunch. If you cure cancer, you will have more cases of neurodegenerative disease. If you cure neurodegenerative disease, a major plague will come for people who are a hundred years old. There is no ultimate solution, nor should there be."

But this reality is blessed with light. "We have to distinguish between life-span and *health*-span," Medzhitov said. "You don't want to live forever, but you do want to be healthy when you're old."

This is what all these inventions and innovations have provided us: a bit more life and a whole lot more comfort as we age. Less pain, anxiety, disabling disease. Less fragility.

As a species, we have strived for immortality and attained only a distant second place. But third place sucked a lot, earlier death, agony.

The Meaning of Jason holds two competing principles in exquisite balance: We must continue to strive, dream, and exercise all the passions that have gotten us this far, while also doing a much better job of accepting death. Death is not just inevitable, not just programmed into us and facilitated in ways by our immune system. It is essential for our survival.

It is not an easy leap to at once be driven by terror of death and yet embrace it with humility and grace. Our continued health lies in creating this balance, as elegant as the balance struck by the immune system itself.

On January 1, 2017, I was back in Colorado, loading the family into the car after a ski day, when my cell phone rang. I figured I'd let it go to voice mail. Big flakes floated down, and I was harried. But the caller was Guy Greenstein, Jason's brother, and I just had this strange feeling.

"Hey, Guy."

"Hey, Matt. I have some bad news. My mom died."

Guy had found their mother collapsed outside of her bathroom. It looked like it had been the heart, and fast.

"The coroner, Mike, he said to me: Didn't I just see you?"

Rest in peace, Catherine Greenstein.

Six months later, I lost my beloved grandmother, Anne Richtel, just a few days shy of her hundredth birthday.

In October of 2017, Ron Glaser went into hospice in the memory care unit. He became confined to a wheelchair because of the risk of falling. He understood little.

"I can put my face literally two inches from his, and he will look through me," Jan Kiecolt-Glaser told me. In her particular spirit, she managed to find the glitter. "There are still golden moments when he recognizes me and smiles."

Two months later, almost a year to the day that Cathy died, Dr. Ben Barres, the guru of dementia and the immune system, passed away on December 27, 2017. He was sixty-three. He'd hoped, he told me, to have an immunotherapy reprieve like Jason. He did leave a monstrous legacy that may yet spare us some of dementia's cruel ignominy. He embodied proof of the value of diversity. Born a woman, he became a man and experienced the world through different eyes, perhaps allowing him to see what others could not.

During this project, death came and went. This, as I say, is not the place I expected to end up. I thought I'd tell the story of Jason driving in the foul-smelling Windstar to Denver to get his injection, waking up one morning to hear his girlfriend say his tumors disappeared, and then going on to another adventure. I thought he'd populate the world with yarns and fumes, pausing only briefly to fuel up at 7-Eleven for more snacks. In that survival story, I imagined, would be hope for all of us.

Not infrequently, after Jason's death, and well before, I began to think of him in a new, and particular, light. I saw him as a son— the son who lost his father. This perspective is excruciating for me because my own son, Milo, is ten. Like Jason, he's a jock—Prodigy 2 is what Jason called him. I'm Milo's coach, just like Jason's dad coached him. Like Jason and his dad, Milo and I are thick as thieves. Just like most fathers and sons. My daughter, Mirabel, is eight, a creative, funny, loving soul; a child of such magnificence, as is her brother. I dared not dream of having such offspring and somehow did. The prospective horror of leaving behind a son and daughter, a family, or losing one, has become palpable. Like so many, I count my blessings every day. Each day, I count with a little more gratitude. We have a finite time in this Festival of Life. It is beautiful. It hurts.

Thanks to science and wisdom both, we possess more comfort as we age, and knowledge about how our bodies work so as to make better choices. When sickness hits, we will get another year, or two or ten. The Argonauts have given us the miracle of extra days, and when my time is nigh, I'll thankfully gulp every extra minute.

But I've also come to see a different reason for hope. The gifts given to us from human learning have come through extraordinary cooperation, through hard-earned and lucky experimentation— in labs, yes, but also in homes and statehouses, and in the "two steps forward, one step back" of cultural, political, social, and scientific advance. We won't skirt inevitable death, not as individuals. However, when we pull back the lens, the Festival of Life can rage on if we find harmony as a species. Maybe, when it comes to it, I will have been able to give my son and daughter tools to carry with them and bring us all a molecule closer to peace.

After Jason died, I stood at the base of his bed and thanked him for always being kind to the little guy. Depending on the setting, each of us can find ourselves as the little guy or the top dog, as needing or being able to give, as supplicant, friend, bully, or antagonist. Each of us, like microscopic bit players in a larger organism, also has an outsized power to signal cooperation, find harmony, to hasten hostilities or dampen them.

The deep friendship I wound up forming with Jason captures a searing truth instructed by the immune system. We are in this together.

Acknowledgments

One day I was talking about the immune system for this book with Dr. Mike McCune, an accomplished researcher and clinician at the University of California, San Francisco. We'd spent hours talking at various points. I thanked him for the generosity of his time.

He said, "I'm trying to build the world's most conversational immunologist."

I asked what he meant, and he explained that immunology needs a translator, someone to bring these concepts to life and explain them to the public.

Dr. McCune, I hope you feel that your time was invested wisely. This is a wish I have too for the dozens upon dozens of scientists and doctors to whom I owe an incredible debt. This group includes the men and women I wrote about and quoted in the book and many others whose names are not included here but whose time and wisdom proved invaluable to me. Please accept my deepest gratitude for your patience, your good humor, and above all, your scientific work. You have saved, strengthened, and lengthened many lives.

Thanks to Dorsey Griffith for your patient research assistance. Vicki Yates, you've been a godsend on this and other projects.

I am lucky to have found a family at William Morrow. Peter Hubbard, editor and friend, thank you for your humor, bedside manner, and great wisdom. Thanks to Nick Amphlett, ever present, ever able. Huge thanks, as always, to Liate Stehlik, a publisher, friend, and unflappable ship's captain in the rocky book-world seas.

Laurie Liss, my agent and sister slightly removed, much love. Our tree isn't dead yet.

I owe tremendous debt to Douglas Preston, world-class writer and teacher, who took on the role of sounding board and periodic editor for this book. I couldn't have asked for better counsel.

Thank you and love to my wife, Meredith Jewel Barad, the foundation of the whole damn thing, and to Milo and Mirabel, our angels, and Uncle Mort and Pickles, our pets. Thanks, Mom and Dad.

To Dr. Mark Brunvand: You spent hours sharing, explaining, opening your heart, becoming teacher and friend. Thank you for all of that and for a lifetime spent doing the same for so many of your patients, steering the narrows.

To Bob Hoff: I will forever carry your story because it taught me so much about courage. You weathered a brutal period in this country, and of course, in your own health and in the deaths of so many friends. Your dignity blows me away. Thank you for being so open. I find it pitiful that the pockets of discrimination remain, and I hope like hell that this sickness, this autoimmunity of big-otry, will subside before it leads to the catastrophic.

To Linda Segre, I offer three words: grace under pressure. I re-alize it can't be as easy as you make it sometimes look. I'm sure the reader shares my gratitude for your sharing of the challenges of making your own way while struggling with the demons of auto-immunity.

To Merredith Branscombe, please accept a double shot of thanks: you told me your story and you acted as an eagle-eyed aide-de-camp in my journalistic thinking. Your experience as a writer and creator added a layer of insight that elevated this effort. Thank you.

To Jason's family, and to Beth, words are insufficient. You treated me like a brother. I am sorry for the loss of Jason and for Cathy. She was a blast of a human being, funny and warm and most certainly the font of Jason's fire.

Jason.

Every so often, I talk to Jason. It's usually a whisper after Milo, my son, has done something special on the baseball field. "Greenie," I'll say, "I'd have called you about that one." Or "Did you see that, Greenie?"

You remain in my heart. I count among my blessings that we came to call each other friend. Your light shines on.

Index

Note: Page references in *italics* indicate photographs.

MATT RICHTEL is a reporter for the *New York Times*. He received the Pulitzer Prize for National Reporting for a series of articles about distracted driving that he expanded into his first nonfiction book, *A Deadly Wandering*. A *New York Times* bestseller, *A Deadly Wandering* tells an intimate story of a fatal car crash and was named a best book of the year by the *San Francisco Chronicle*, *Christian Science Monitor*, *Kirkus Reviews*, *Winnipeg Free Press*, and Amazon. Richtel has appeared on NPR's *Fresh Air* and *PBS NewsHour*, and in other major media outlets. He lives in San Francisco, California.

MORE FROM MATT RICHTEL

A Deadly Wandering

One of the decade's most original and masterfully reported books, *A Deadly Wandering* by Pulitzer Prize–winning *New York Times* journalist Matt Richtel interweaves the cutting-edge science of attention with the tensely plotted story of a mysterious car accident and its aftermath to answer some of the defining questions of our time: What is technology doing to us? Can our minds keep up with the pace of change? How can we find balance?

An Elegant Defense

A grand tour of the human immune system and the secrets of health, by the Pulitzer Prize–winning *New York Times* journalist.

Drawing on his groundbreaking reporting for the *New York Times* and based on extensive new interviews with dozens of world-renowned scientists, Matt Richtel has produced a landmark book, equally an investigation into the deepest riddles of survival and a profoundly human tale that is movingly brought to life through the eyes of his four main characters, each of whom illuminates an essential facet of our "elegant defense."